Workload Dependent Mitigation Approaches for Performance Variability

Ji-Yung Lin • Michalis Noltsis • Dimitrios Soudris •
Francky Catthoor

Workload Dependent Mitigation Approaches for Performance Variability

Ensuring Timing Guarantees of Integrated Circuits

Springer

Ji-Yung Lin
IMEC
Leuven, Belgium

Michalis Noltsis
Zografou Campus
National Technical University of Athens
Athens, Greece

Dimitrios Soudris
Zografou Campus
National Technical University of Athens
Athens, Greece

Francky Catthoor
IMEC
Leuven, Belgium

ISBN 978-3-031-85623-5 ISBN 978-3-031-85624-2 (eBook)
https://doi.org/10.1007/978-3-031-85624-2

This Springer imprint is published by the registered company Springer Nature Switzerland AG
The registered company address is: Gewerbestrasse 11, 6330 Cham, Switzerland

Preface

Performance variability, the phenomenon that the execution time of the same application varies every time it is executed, is ubiquitous in modern computers. Performance variability is especially detrimental to time-sensitive and time-critical applications, which are the applications required to be completed before a certain deadline. With continuous development toward smaller devices and more dynamic architectures and software, performance variability increases and threatens the reliability of both time-sensitive and time-critical applications.

To reduce performance variability, real-time mitigation approaches are used in the online operation of various modern systems and platforms, including processors, memory organizations, and communication networks. Mitigation approaches use on-chip monitors to detect the amount of variability at run time. Based on the amount of variability and the prediction of future workload, a run-time controller changes the speed of the system or the platform by altering system knobs. Among all kinds of mitigation approaches, workload-dependent approaches operate on the principle of using the design-time profiling information to limit and reduce the uncertainties in future execution time calculation at run time. Therefore, only workload-dependent approaches can ensure full timing guarantees, so they are the most promising option for tackling performance variability while minimizing system costs.

This book elaborates on the concept and realization of a workload-dependent mitigation approach tackling various performance variability. We have improved the ways of tackling performance variability from the architecture and software levels based on two features, heterogeneous datapaths (HDP) and the concept of dynamic scenarios. Furthermore, we have developed a cross-layer reliability and variability mitigation framework to simultaneously ensure functional correctness and timing guarantees.

A full timing-guaranteed approach is necessary to ensure the reliability of systems. This is especially crucial in highly scaled technology nodes, where the intrinsic variability at both fabrication-time (static part) and use-time (time-dependent part) are increasing significantly compared to conventional nodes. With the contributions of this book, we look forward to the development of a general performance variability mitigation approach which ensures full timing guarantees for highly reliable products even in the most advanced technology nodes.

The completion of this book is the result of the collective efforts of many brilliant and generous minds across various groups. We would like to extend our sincere thanks to the following individuals for their invaluable contributions to the research and the book: Pieter Weckx, Ben Kaszer, Subrat Mishra, Alessio Spessot, Odysseas Zografos, and Dwaipayan Biswas of IMEC; Prof. Guido Groeseneken and Prof. Georges Gielen of KU Leuven; and Nikolaos Zompakis and Dimitrios Rodopoulos of NTUA. Finally, we would like to give special thanks to our families for their unwavering support throughout the research and writing process.

Leuven, Belgium Ji-Yung Lin
Athens, Greece Michalis Noltsis
Athens, Greece Dimitrios Soudris
Leuven, Belgium Francky Catthoor

Declarations

Competing Interests The authors have no competing interests to declare that are relevant to the content of this manuscript.

Contents

Acronyms

AI	Artificial Intelligence
ALU	Arithmetic Logic Unit
BE	Best-Effort
BEOL	Back-End-of-Line
BPP	Bit Per Pixel
BTI	Bias Temperature Instability
CMOS	Complementary Metal-Oxide-Semiconductor
CRR	Checkpointing and Rollback-Recovery
CSR	Control and Status Register
DC	Direct Current
DEMUX	Demultiplexer
DMA	Domain-Specific Accelerator
DMR	Dual Modular Redundancy
DS	Dynamic Scenario
DVF	Deadline Vulnerability Factor
DVFS	Dynamic Voltage and Frequency Scaling
ECC	Error Correction Circuitry
EM	Electromigration
ERW	Error Rate Wall
FEOL	Front-End-of-Line
GPU	Graphic Processing Unit
HCI	Hot Carrier Injection
HDP	Heterogeneous Datapath
HW	Hardware
ICG	Integrated Clock Gating
IPC	Instruction Per Cycle
ISA	Instruction Set Architecture
JPEG	Joint Photographic Experts Group
LER	Line Edge Roughness
MCU	Minimum Coded Unit
MEOL	Middle-End-of-Line

MGG	Metal Grain Granularity
MOSFET	Metal-Oxide-Semiconductor Field-Effect Transistor
MP2	MPEG-1 Audio Layer II
MPEG	Moving Picture Experts Group
MUX	Multiplexer
NBTI	Negative Bias Temperature Instability
OS	Operating System
P&R	Place and Route
PBTI	Positive Bias Temperature Instability
PC	Personal Computer
PG	Power Gating
PID	Proportional–Integral–Derivative
PLL	Phase-Locked Loop
PVF	Performance Vulnerability Factor
QCIF	Quarter Common Intermediate Format
RAS	Remote Access Service
RDF	Random Dopant Fluctuation
RFID	Radio-Frequency Identification
RGG	Ratio of the Average Grain Size to the Gate Area
RTL	Register-Transfer Level
RTN	Random Telegraph Noise
RTS	Run-Time Situation
SoC	System-on-Chip
SRAM	Static Random-Access Memory
SW	Software
TDDB	Time-Dependent Dielectric Breakdown
TDP	Thermal Design Budget
TMR	Triple Modular Redundancy
TN	Thread Node
VLSI	Very Large-Scale Integration
WCET	Worst-Case Execution Time
WD	Workload-Dependent
WFV	Work Function Variation

Chapter 1
Introduction

Abstract This chapter introduces the research topic of this book, including the background, motivation, scope, and significance of the workload-dependent mitigation approaches. We will start with an introduction of the goal of this work, the timing guarantee requirements of applications. Then, we will elaborate on performance variability in integrated systems, which is the main threat to timing guarantees. After that, the main gaps in the literature in the field of timing guarantees are discussed. Finally, we will give an overview of the book, including the problem statement, our main contributions in filling the gaps of the state-of-the-art, and the structure of the whole book.

1.1 Timing Guarantees of Applications

With the rapid development of the semiconductor transistor technology, electronic devices made of Very Large-Scale Integration (VLSI) chips become necessities in our daily lives, with countless functions such as computation, communication, entertainment, etc. Ever since the early age of VLSI chips in the 1970s, the number of transistors per area approximately doubled every year, following Moore's Law [11] as depicted in Fig. 1.1. In the modern era, a typical chip can contain billions of transistors. The continuous transistor downscaling has fueled great improvement in the chip performance and cost. Over the past 40 years, the performance has grown by 50,000× [7], and the price per transistor has been cut by 1,000,000× [8], enabling wide ranges of applications from the tiny RFID tags for object identification to the gargantuan supercomputers driving scientific computations such as weather forecasts. The huge number of transistors also enables complicated system designs. A highly integrated system-on-chip (SoC) design can contain hundreds of processor cores, multiple levels of memory, and sophisticated interconnection networks all inside a single chip.

Even with the high transistor counts and system complexity, VLSI chips are required to ensure reliability throughout the whole lifespan. The VLSI chips are required to operate without failures regardless of the changeable internal and

J.-Y. Lin et al., *Workload Dependent Mitigation Approaches for Performance Variability*, https://doi.org/10.1007/978-3-031-85624-2_1

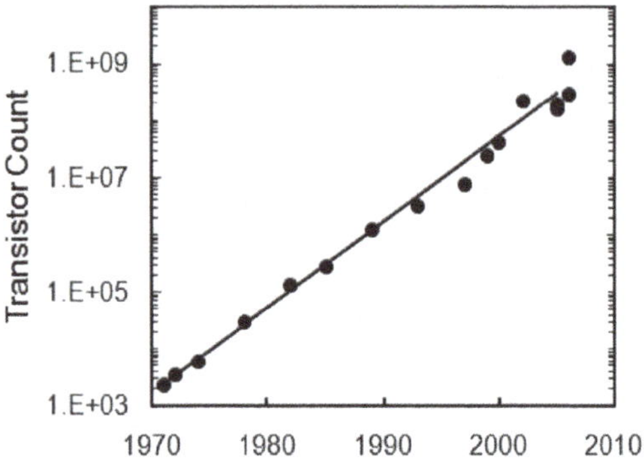

Fig. 1.1 Intel CPU transistor count trend. Replotted from [3]

Fig. 1.2 General requirements of chip reliability

external conditions. The requirements of chip reliability are depicted in Fig. 1.2. In general, the following are the reliability requirements for a typical chip:

- **Functional correctness**: The chip needs to operate without failing and produce correct computation results under all circumstances.
- **Timing guarantees**: The chip needs to ensure the timeliness of delivering the computation results. The results could be obsolete if not produced in time and thus fail to serve the purpose they are designed for.
- **Thermal constraints**: The chip temperature needs to be limited under a certain level even though the devices and interconnects dissipate heat. Otherwise, thermal runaway could happen and the devices and interconnects could break down due to high temperature.
- **Parametric correctness**: Beyond the aforementioned hard reliability requirements, there are other parametric requirements for different metrics specific to each field. For example, energy consumption is limited for battery-powered and energy-harvesting platforms.
- **Lifetime**: The aforementioned requirements need to be ensured for the whole lifetime specification even though the chip could wear out. Normally consumer

products need to be reliable for 5 years, while safety-critical products need to work for at least 10 years.

Timing guarantee is the main focus of this book among all reliability requirements. Among the diverse applications of VLSI chips, most applications care about the timeliness of delivering the calculation results. In practice, results are required to be calculated before certain deadlines. Failing to meet the deadline constraints will result in numerous consequences from quality loss to safety concerns. Based on the criticality of ensuring timing guarantees, applications can be further classified into three groups:

- **Time-critical**: Timeliness is vital for the function of these applications, where they must always meet their timing specifications without any deadline violations. For these applications, missing deadlines might cause catastrophic outcomes such as safety problems. Vehicles, avionics, and biomedical systems are examples of time-critical applications. Since these systems closely interact with the real-world environment and physical bodies, delayed computation results mean that the results are based on outdated information about the real world, causing dysfunction which could cost human lives.
- **Time-sensitive**: Deadline violations cause quality loss for these applications. Occasional deadline misses are tolerable but must be guaranteed to be below a specified threshold. Hence, the variability and reliability management system should be deployed to monitor whether this timing specification is fulfilled and proactively prevent potential violation of the specification. Multimedia such as real-time video is one example of this type. The timing of a video is specified by its frame rate, which defines the frequency at which consecutive images (i.e., frames) are displayed on the screen. In real-time video processing, failing to present a frame in time occasionally is acceptable, since it may not be noticeable by human eyes. However, losing several frames in a sequence can cause visible video lagging, significantly degrading user experiences, which should be prevented.
- **General**: Even though without obvious timing specifications, nearly all the remaining applications have implicit maximum latency constraints, in consideration of user experiences. For instance, users of consumer products (e.g., mobile phones or PCs) are not willing to wait longer than a certain time for a response from their application running on their processing platform. Thus, the latency violation rate should stay below a small threshold, which has to be guaranteed for products to be economically viable in the market.

In conclusion, timing requirements are prevalent in every application domain. Therefore, ensuring timing guarantees is a crucial property for VLSI chips to achieve high quality and reliability of their functions. Moreover, other reliability requirements and quality-of-service (QoS) constraints (e.g., chip area) can exist on top of timing requirements, which further complicates the problem. Nevertheless, the approaches for tackling different reliability issues generally follow similar principles. In those cases, the reliability constraints can be generalized to a function in a multidimensional Pareto space. In this book, we will also discuss how the main principles of our approaches can be extended to include other reliability requirements and QoS constraints.

1.2 Performance Variability in Integrated Systems

Despite the importance of timing guarantees, the execution time of applications in modern VLSI chips tends to be dynamic and unpredictable. The execution time of an application can be different every time it is executed, due to numerous variability sources. The term "performance variability" is used to describe the variations in execution time, which poses serious threats to timing guarantees.

As an illustrative example of performance variability, Fig. 1.3 shows the dynamism in the execution time of an MPEG-1 Audio Layer II (MP2) decoding application for decoding bitstream frames of Vivaldi's "Four Seasons" Concerti, in which each data point represents the execution time of decoding a bitstream frame of the audio stream. The measured execution time for processing each frame is highly diverse, ranging from below 1 second to above 2 seconds.

Performance variability can be originated from all abstraction layers from the device level to the application level. Figure 1.4 instantiates typical performance variability sources in each level. The timescale of each variability source is widely diverse. In general, the higher the level, the larger the timescale of a single variability event. For example, at the lowest device level, the delay jitter of a logic gate caused by bias temperature instability (BTI) can be even smaller than 1 picosecond [9]. At the circuit level, the glitches in the voltage supply can oscillate with a period smaller than 1 nanosecond [12]. At the microarchitecture level, whether a branch prediction hits or misses results in tens of cycles difference in executed cycles [4], equivalently tens of nanoseconds on a 1-GHz processor core. At the highest application level,

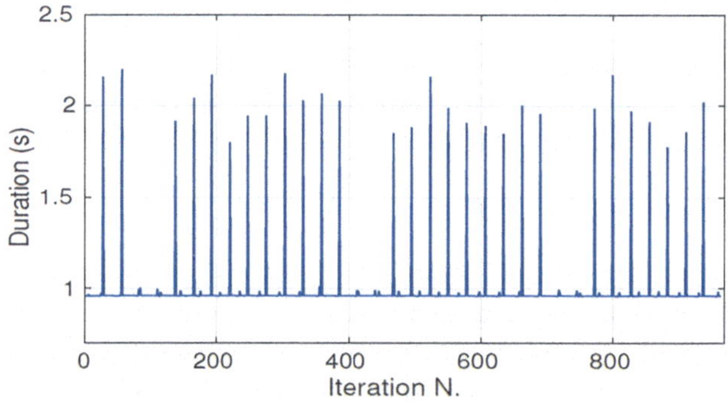

Fig. 1.3 Dynamism in execution time of an MP2 decoding application for decoding of Vivaldi's "Four Seasons" Concerti. Each data point represents the execution time of decoding a bitstream frame of the audio stream

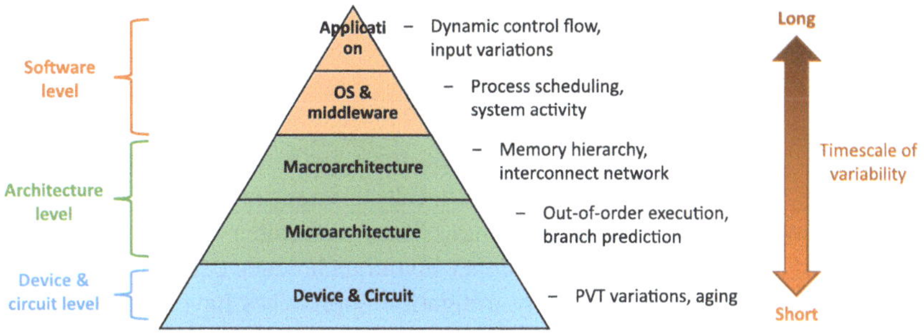

Fig. 1.4 Main sources of performance variability in every abstraction layer

different input data can result in seconds of difference in execution time, as the example of Fig. 1.3.

With the endless pursuit of technology scaling and system integration, performance variability is increasing at all levels. At the device level, the downscaling of device dimensions makes the fabrication process more and more difficult to control, which increases variability among devices [6, 10]. At the architecture level, the increase in transistor count enables more and more dynamic, complicated, and heterogeneous architectures, which increases variability in execution time [5, 13, 16, 17]. At the software level, the trends toward dynamic execution flow and distributed computation also introduce more variables in execution time, which exacerbates performance variability [15, 17].

Therefore, mitigation approaches for performance variability are indispensable to ensure timing guarantees. However, with various variability sources from all abstraction layers, a holistic solution to performance variability is no easy task. Moreover, interdependencies can be present across multiple layers, further complexing the problem. Hence, the main topic of this book is to overcome these challenges to develop solutions to performance variability.

1.3 Main Gaps in Related Research

Though approaches for ensuring timing guarantees have been developed and used over the past years, advances in technology and computer architecture have brought new challenges to timing guarantees. Therefore, we suppose there is still plenty of room for improvement in the state-of-the-art approaches. In the following, we briefly introduce the important gaps in the related field, while the details will be discussed in later chapters.

- Calculation of the upper bound of execution time is the key to ensuring timing guarantees for any approach. In the past, Worst-Case Execution Time (WCET) [1, 17] analysis was developed to compute the upper bound of the execution time of an application under all possible variability conditions. While WCET is the foundation of the conventional approaches of timing guarantees, the complexity of modern hardware and computer architectures greatly increases the execution time overhead induced by WCET to an unacceptable amount, making WCET-based approaches undesirable. Since then, the specific problem formulation of mitigating performance variability with full timing guarantees has not been achieved up to now in state-of-the-art. Many researchers have claimed that only static and highly pessimistic upper bound techniques can ever accomplish this. In this context, though many mitigation approaches have been proposed, their guarantees of timing assurance or deadline miss rate are questionable [1]. Nevertheless, we suppose that the conjecture of no possible timing guaranteed approach is too pessimistic and that timing guarantees can be ensured by run-time workload-dependent approaches, so a paradigm shift toward full timing guarantee can be created. More details are discussed in Chap. 3.

- To ensure timing guarantees, real-time scheduling is the most prominent approach [2]. However, the research on real-time scheduling is normally based on simplified mathematical processor models, which cannot fully reflect the performance variability below the software level. Therefore, the state-of-the-art approaches are insufficient in dealing with lower level variability. Conventional WCET-based and best-effort mitigation approaches are not sufficient to fill this gap since they are not real time or online. Since performance variability comes from each abstraction layer in different timescales, it is necessary to establish a cross-layer framework which combines all techniques involved in the performance variability of each layer. In this way, timing guarantees can be holistically ensured, and mitigation costs can be accurately evaluated. More details are discussed in Chap. 7.
- As far as other reliability requirements (e.g., functional correctness) are concerned, there is a trend of integrating multiple techniques to deal with multiple reliability objectives in the recent development of mitigation approaches. For example, real-time scheduling can be combined with checkpointing and rollback-recovery, as well as thermal management to tackle three reliability objectives simultaneously [14]. While these techniques extend the scope of mitigation approaches for reliability, a holistic framework is required to solve the inter-dependencies of all these objectives across all abstraction layers. This kind of framework can be created to pave the way for a unified methodology for all reliability objectives. More details are discussed in Chap. 7.

1.4 Problem Statement

This research aims to develop a methodology for ensuring timing guarantees in modern systems and platforms of integrated circuits. The conventional WCET-based approaches for ensuring timing guarantees are based on imposing a total execution time margin of the worst-case conditions of all performance variability sources. However, with the growing complexity of modern computer architecture and applications, as well as the increasing variability in advanced technology nodes such as voltage variations, the execution time margin considered in WCET becomes unacceptably high, inducing too much overhead in performance and energy. Hence, new methodologies are required to ensure timing guarantees for state-of-the-art processors and applications. We suppose that run-time workload-dependent mitigation approaches, which can adapt to every run-time situation and only add the required overhead to meet the timing requirements, can be the new methodology that ensures timing guarantees with low overhead. Therefore, the goal of this research is to realize such a methodology.

In terms of the target hardware, this work focuses on a methodology on processor cores, especially datapaths. Nevertheless, we have derived generic principles required to realize a full timing-guaranteed platform, with which the methodology on processors is developed. Therefore, the extension to other hardware, such as communication networks and memory organization, is possible based on the same principles.

1.5 Contributions of This Book

We summarize the contributions of this book to the state-of-the-art as follows:

- We discuss the concepts of system scenarios, adaptive scenarios, and dynamic scenarios. Based on system and adaptive scenarios, a closed-loop controller that manages performance variation and ensures timing deadlines through proper Dynamic Voltage and Frequency Scaling (DVFS) actuations is introduced and instantiated on a target platform where its capabilities and limitations are explored.
- We have developed a novel workload-dependent multi-timescale mitigation approach, which can ensure full timing guarantees in the presence of architecture-level and software-level performance variability of different timescales down to the sub-millisecond level, while the energy consumption is minimized at the same time. This approach aims to mitigate performance variability for timing guarantees of all application types (time-critical, time-sensitive, and general). This approach has two main features that improve the state-of-the-art approaches. First, the heterogeneous datapath (HDP) knob enables fast switching between different processor speeds, so that the performance variability can be tackled in sub-millisecond granularity. Second, on the principle of dynamic scenarios, the proposed approach can derive the upper bound of execution time at run-time to prevent deadline misses.
- To evaluate the mitigation approach, we have developed a cycle-accurate simulation flow, which can simulate the performance variability in cycle-level granularity. The flow enables the benchmark of important processor metrics

(execution time, energy consumption, etc.) to compare the reliability and costs achieved by different performance variability mitigation approaches.

- We develop a cross-layer reliability and variability mitigation framework, which can tackle the device-level and circuit-level performance variability to ensure both functional correctness and timing guarantees. The framework combines mitigation approaches of three different layers, including mitigation approaches of device-level variability (e.g., voltage variations) at the circuit level, checkpointing and rollback-recovery at the register level, and performance variability mitigation at the processor level.
- With the aforementioned framework, an analysis of cross-layer dependencies and the holistic system impact of variability is performed. Co-simulation across layers shows the cost and the system reliability limit in ensuring the dual reliability requirements. The simulation results reveal the phenomenon of the "error rate wall" (ERW), a system-specific error rate limit beyond which reliability cannot be ensured. In addition, we show that the reliability of the overall system can be improved by changing system parameters, which indicates the potential of optimizing system parameters to enhance reliability.

1.6 Structure of This Book

In Chap. 2, we give an overview of variability in integrated circuits, in which every source of performance variability in each layer is addressed. Moreover, we discuss the general principles of the approaches to tackling variability, including guardbanding and run-time mitigation. In Chap. 3, we elaborate on the state-of-the-art mitigation approaches for performance variability, in which each component (monitors, system knobs, and adaptive control algorithms) is discussed in detail. In Chap. 4, we present the concept of system and adaptive scenarios, as well as deploy a PID mitigation scheme on a fine workload granularity, applying Dynamic Voltage and Frequency Scaling (DVFS) actuations. In Chaps. 5 and 6, we present the workload-dependent multi-timescale approach we have developed for tackling performance variability based on the concept of dynamic scenarios. Chapter 5 introduces the theoretical basis and the design of this approach, while Chap. 6 shows the simulation flow and the results. Thereafter, we propose the cross-layer reliability and variability mitigation framework in Chaps. 7 and 8. The principles and theoretical basis are described in Chap. 7, and the simulation results are discussed in Chap. 8. Finally, Chap. 9 summarizes the main messages and contributions of this book and discusses the future perspective of this work (Fig. 1.5).

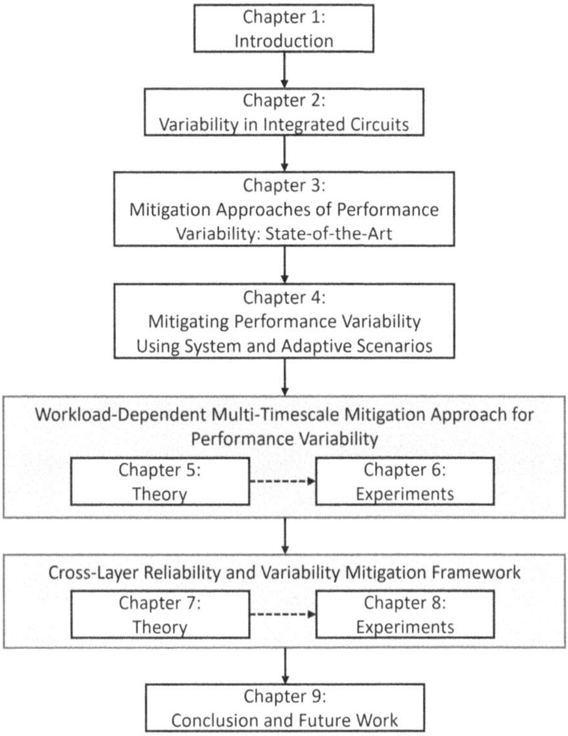

Fig. 1.5 Structure of the book and overview of the chapters

References

1. Abella, J., Hernandez, C., Quiñones, E., Cazorla, F.J., Conmy, P.R., Azkarate-askasua, M., Perez, J., Mezzetti, E., Vardanega, T.: WCET analysis methods: pitfalls and challenges on their trustworthiness. In: 10th IEEE International Symposium on Industrial Embedded Systems (SIES), pp. 1–10 (2015). https://doi.org/10.1109/SIES.2015.7185039
2. Bambagini, M., Marinoni, M., Aydin, H., Buttazzo, G.: Energy-aware scheduling for real-time systems: a survey. ACM Trans. Embed. Comput. Syst. **15**(1), 1–34 (2016). https://doi.org/10.1145/2808231
3. Bohr, M.: The evolution of scaling from the homogeneous era to the heterogeneous era. In: 2011 International Electron Devices Meeting, pp. 1.1.1–1.1.6 (2011). https://doi.org/10.1109/IEDM.2011.6131469
4. Eyerman, S., Smith, J., Eeckhout, L.: Characterizing the branch misprediction penalty. In: 2006 IEEE International Symposium on Performance Analysis of Systems and Software, pp. 48–58 (2006). https://doi.org/10.1109/ISPASS.2006.1620789
5. Furber, S.: Microprocessors: the engines of the digital age. Proc. R. Soc. A: Math. Phys. Eng. Sci. **473**, 20160893 (2017). https://doi.org/10.1098/rspa.2016.0893
6. Grasser, T., Kaczer, B., Goes, W., Reisinger, H., Aichinger, T., Hehenberger, P., Wagner, P.J., Schanovsky, F., Franco, J., Roussel, P., Nelhiebel, M.: Recent advances in understanding the bias temperature instability. In: 2010 International Electron Devices Meeting, pp. 4.4.1–4.4.4 (2010). https://doi.org/10.1109/IEDM.2010.5703295
7. Hennessy, J.L., Patterson, D.A.: Computer Architecture, 6th edn. A Quantitative Approach, 6th edn. Morgan Kaufmann Publishers Inc., San Francisco, CA, USA (2017)

8. Hutcheson, G.D.: Moore's law, lithography, and how optics drive the semiconductor industry. In: Goldberg, K.A., Felix, N.M. (eds.) Extreme Ultraviolet (EUV) Lithography IX. *Society of Photo-Optical Instrumentation Engineers (SPIE) Conference Series*, vol. 10583, p. 1058303 (2018). https://doi.org/10.1117/12.2308299
9. Kaczer, B., Mahato, S., Valduga de Almeida Camargo, V., Toledano-Luque, M., Roussel, P.J., Grasser, T., Catthoor, F., Dobrovolny, P., Zuber, P., Wirth, G., Groeseneken, G.: Atomistic approach to variability of bias-temperature instability in circuit simulations. In: 2011 International Reliability Physics Symposium, pp. XT.3.1–XT.3.5 (2011). https://doi.org/10.1109/IRPS.2011.5784604
10. Kuhn, K.J., Giles, M.D., Becher, D., Kolar, P., Kornfeld, A., Kotlyar, R., Ma, S.T., Maheshwari, A., Mudanai, S.: Process technology variation. IEEE Trans. Electron Devices **58**(8), 2197–2208 (2011). https://doi.org/10.1109/TED.2011.2121913
11. Moore, G.E.: Cramming more components onto integrated circuits. Electron. Mag. **38**(8), 114–117 (1965)
12. Muhtaroglu, A., Taylor, G., Rahal-Arabi, T.: On-die droop detector for analog sensing of power supply noise. IEEE J. Solid State Circuits **39**(4), 651–660 (2004). https://doi.org/10.1109/JSSC.2004.825120
13. Patterson, D.: 50 years of computer architecture: From the mainframe CPU to the domain-specific TPU and the open RISC-V instruction set. In: 2018 IEEE International Solid—State Circuits Conference—(ISSCC), pp. 27–31 (2018). https://doi.org/10.1109/ISSCC.2018.8310168
14. Ranjbar, B., Hosseinghorban, A., Salehi, M., Ejlali, A., Kumar, A.: Toward the design of fault-tolerance-aware and peak-power-aware multicore mixed-criticality systems. IEEE Trans. Comput. Aided Des. Integr. Circuits Syst. **41**(5), 1509–1522 (2022). https://doi.org/10.1109/TCAD.2021.3082495
15. Skinner, D., Kramer, W.: Understanding the causes of performance variability in HPC workloads. In: IEEE International. 2005 Proceedings of the IEEE Workload Characterization Symposium, 2005, pp. 137–149 (2005). https://doi.org/10.1109/IISWC.2005.1526010
16. Taylor, M.B.: Is dark silicon useful? harnessing the four horsemen of the coming dark silicon apocalypse. In: DAC Design Automation Conference 2012, pp. 1131–1136 (2012)
17. Wilhelm, R., Engblom, J., Ermedahl, A., Holsti, N., Thesing, S., Whalley, D., Bernat, G., Ferdinand, C., Heckmann, R., Mitra, T., Mueller, F., Puaut, I., Puschner, P., Staschulat, J., Stenström, P.: The worst-case execution-time problem–overview of methods and survey of tools. ACM Trans. Embedded Comput. Syst. **7**(3) (2008). https://doi.org/10.1145/1347375.1347389

Chapter 2
Variability in Integrated Circuits

Abstract In this chapter, we elaborate on a wide variety of general variability sources in all abstraction layers. Each variability source can affect many quality metrics, such as speed, power, reliability, etc. Nevertheless, we focus on how these variability sources would cause performance variability in integrated systems. Furthermore, we discuss the general strategies used in tackling variability.

2.1 Device- and Circuit-Level Variability

Performances of devices and circuits of the same design are intended to be identical from one to another and constant over the whole lifetime. However, due to numerous nonideal effects, their performances are actually nonuniform and change over time. Due to process variation [48], differences already exist among devices and circuits when just fabricated, known as *time-zero variability*. Then, when in operation, their performance will change over time, which is called *time-dependent variability*. The variability can be influenced by the workload, environmental conditions, or the stochastic nature of devices.

Figure 2.1 shows the taxonomy of the main sources of variability. Time-zero variability is attributed to process variation. And for time-dependent variability, the main sources are voltage variation, temperature variation, and aging effect. In the next few subsections, we elaborate on the main sources of device- and circuit-level variability.

2.1.1 Process Variation

A typical VLSI chip contains billions of transistors which all have to operate correctly. Among all these transistors, no two devices are identical, even at the time they are just manufactured. The differences mainly come from process variation, which are the variations originated from nonidealities in process control or the stochastic nature of matter. Composed of these variable devices, the final products

J.-Y. Lin et al., *Workload Dependent Mitigation Approaches for Performance Variability*, https://doi.org/10.1007/978-3-031-85624-2_2

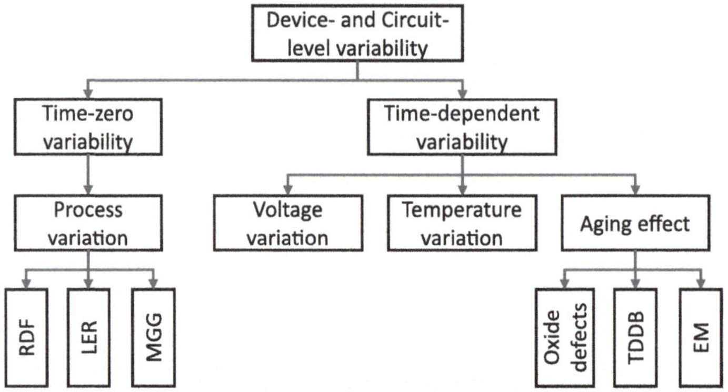

Fig. 2.1 Taxonomy of the sources of device- and circuit-level variability. Only the main sources are enumerated and the miscellaneous sources are left out for clarity

Fig. 2.2 The trend of the average number of dopant atoms reduced with technology nodes. (Replotted from [39])

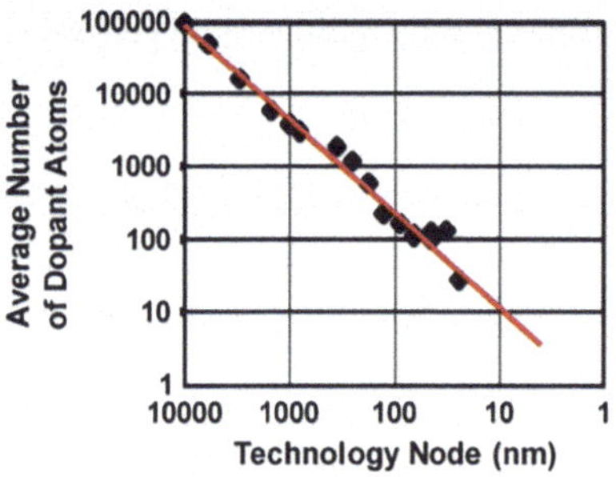

are consequently nonuniform in speed, power, and any other quality metrics, which directly affects the yield. Therefore, process variation has become one of the main design challenges in the semiconductor industry, since it is directly linked to the design and manufacturing, and determines to some extent the final cost of the products.

In the following, we discuss some of the most prominent sources of process variations:

- **Random Dopant Fluctuation**: Random Dopant Fluctuation (RDF) is the difference in the number and the distribution of dopants in the channel of devices, due to the stochastic process of implant ionization [58]. It arises as a prominent variation source as a consequence of the downscaling of transistor dimensions, in that the average number of dopant atoms decreases as a function of technology, as shown in Fig. 2.2 [39]. In the advanced technology nodes, the average dopants per transistor can be fewer than 100, resulting in high randomness in the positions and number of dopants. Consequently, there will be significant variability in many important device parameters, such as threshold voltage (V_{th}), subthreshold slope (SS), cutoff frequency (f_t), and gate capacitance (C_g). In V_{th} variability, RDF is one of the major contributors [38].

By atomistic simulations, an empirical relationship for planner devices with channel length below 100 nm and dopant concentration above $10^{18}\,\text{cm}^{-3}$ is as below [3]:

$$\sigma V_{\text{th}} = 3.18 \times 10^{-8} \frac{T^{\text{OX}} N_{0.4}^{\text{tot}}}{\sqrt{L_{\text{eff}} W_{\text{eff}}}} \tag{2.1}$$

in which T^{OX}, N_{tot}, L_{eff}, W_{eff} represent the oxide thickness, total doping concentration, effective length, and effective width, respectively. The relationship shows a polynomial dependence of V_{th} variability on the total doping concentration.

- **Line Edge Roughness**: Line Edge Roughness (LER) is the roughness and thickness variations of patterned structures, which subsequently result in randomly varied device dimensions such as gate length and width. It is caused by the limitations and nonidealities of both the sub-wavelength lithography and etching process [58]. LER consequently induces both the increase of the leakage current and the variation of V_{th}.

  The amount of LER is typically expressed by its 3σ value [3], where σ represents the standard deviation of the distances from the average edge to the actual edge at different local positions. Measurements of 3σ values from various labs are shown in Fig. 2.3 [3], which indicates that LER is not scaled proportionally with the downscaling of line widths by the advance of lithography processes. This results in an increase of V_{th} variability by LER in advanced technology nodes.
- **Metal Grain Granularity**: Metal Grain Granularity (MGG) is the variability in grain boundaries of transistor gates, which are made of metal or polysilicon [17, 58]. Depending on the process conditions and the randomness in the process of crystallization, the grains can have different sizes and crystal directions, resulting in work function variation (WFV) [71] in the gate area. Consequently, local V_{th} conditions are created, and the aggregated V_{th} varies from device to device [12].

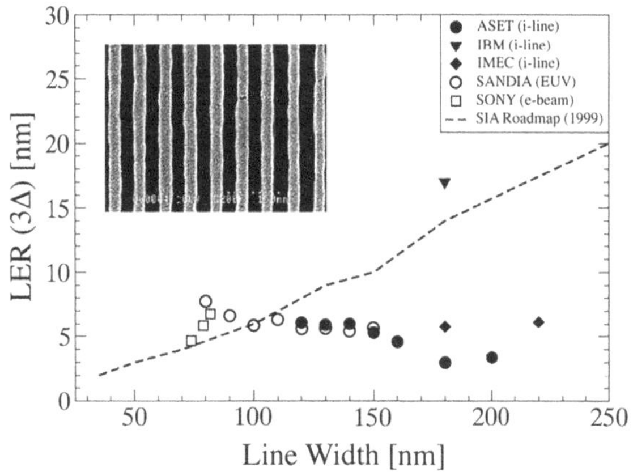

Fig. 2.3 LER measurements for advanced lithography processes reported by various labs. The dashed line indicates the required levels in Semiconductor Industry Association (SIA) roadmap. The inset shows LER observed in sub-100-nm e-beam generated lines. (Replotted from [3])

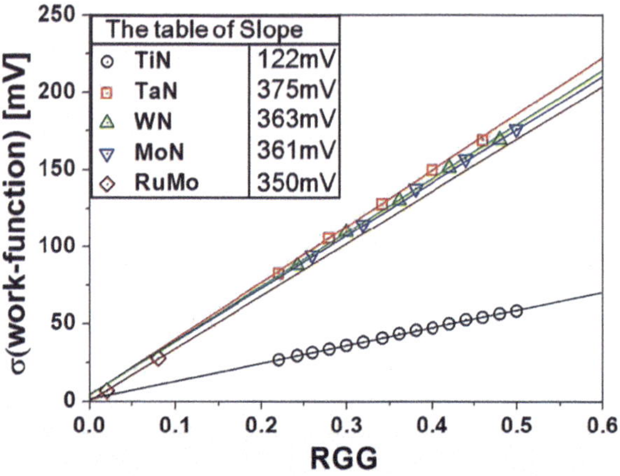

Fig. 2.4 Relations of WFV and RGG for various gate materials (TiN, TaN, WN, MoN, and RuMo [49]). (Replotted from [47])

MGG-induced V_{th} variability scales according to the *ratio of the average grain size to the gate area* (RGG) [47]. Figure 2.4 shows the relations of WFV and RGG for various gate materials [47] (TiN, TaN, WN, MoN, and RuMo [49]). With the same gate material and grain size, the WFV is increased with the shrinking of the gate area. This trend indicates that the V_{th} variability by MGG is increased with the gate area reduction by technology scaling.

2.1.2 *Voltage Variation*

While devices and circuits are intended to function at a stable supply voltage level, they actually suffer from voltage variations occasionally during operation. The main sources of voltage variations include the di/dt noise and the IR drop [45], which originate from parasitic inductances and resistances in the power distribution network, respectively.

The variation in supply voltage changes the switching current of devices, consequently impacting the propagation delay of logic gates and the speed of the entire circuit. Typically, the relation of the propagation delay (t^{pd}), supply voltage (V_{dd}), and threshold voltage (V_{th}) follows the following equation, where α is the velocity saturation index typically ranging from 1 to 2 [30, 56]:

$$t^{pd} \propto \frac{V_{dd}}{(V_{dd} - V_{th})^{\alpha}} \tag{2.2}$$

In prior work, 10% change in V_{dd} is estimated to cause 30% change in t^{pd} in the 130 nm technology, while 1% change in V_{dd} causes 4% t^{pd} in 90 nm technology [50, 67]. Thus, a voltage decrease can significantly enlarge propagation delays, causing logic errors. If frequency scaling is applied, the clock frequency can be changed

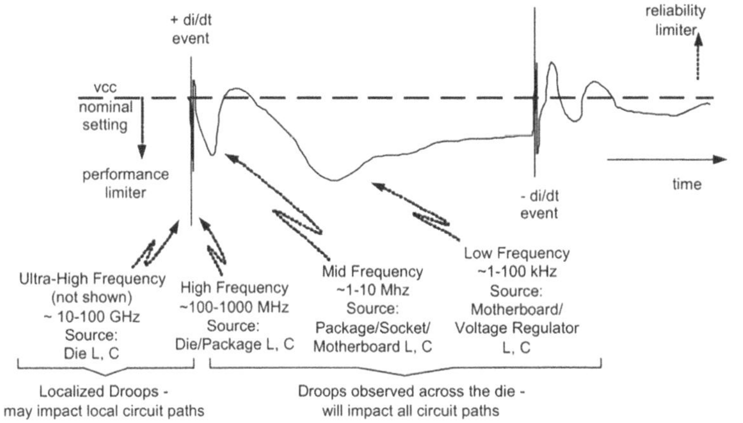

Fig. 2.5 Illustration of a voltage droop occurrence, including all its frequency components. (Replotted from [45])

at run-time to avoid errors. Generally, the maximum achievable clock frequency ($f_{\max}$) of a digital circuit is inversely proportional to the delay of the critical path. Therefore, it follows the equation below [70]:

$$f_{\max} \propto \frac{(V_{\rm dd} - V_{\rm th})^{\alpha}}{V_{\rm dd}} \tag{2.3}$$

Thus, when the voltage is decreased, the clock frequency needs to be reduced, which results in performance variability.

A particularly challenging type of voltage variation is the *voltage droops* [8]. Generated by a sudden increase in the switching activity of the circuit, an abrupt high current is induced to the power delivery network, causing a large IR drop. As shown in Fig. 2.5, a voltage droop contains a wide range of frequency components, in which the high-frequency component occurs in only a few nanoseconds. It is reported that the effect of voltage droops can reach more than 10% voltage reduction and 15% delay increment [8], resulting in logic errors. Due to its nanosecond-level timescale, voltage droops are required to be handled within only a few clock cycles, posing a severe challenge in reliability.

2.1.3 Temperature Variation

As the number of transistors increases, so does the power density, resulting in a rise in chip temperature. If not managed properly, the aggregated power consumption can exceed the thermal design budget (TDP), leading to an overly high temperature which causes chip breakdown. Hence, thermal management and heat dissipation techniques come to the rescue, ensuring the chip's temperature under the thermal constraint.

Besides the average temperature, chips also suffer from uneven distribution of temperatures, which depends on the environment, workload, and heat transportation. Unbalanced power consumption can create local *hotspots*, resulting in performance variation across the chip, such as nonuniform speed among the cores in a multicore processor.

Temperature fluctuation also influences circuit performance. On the one hand, device performance has a dependency on temperature. On the other hand, increased temperature leads to increased metal resistance, lengthening the propagation delay of interconnects. Typically, the temperature fluctuates at a slower timescale of the hundreds of microsecond level [6], due to the relatively large time constant of heat transportation. Therefore, it can be mitigated by a wide variety of techniques, such as DVFS [10], task mapping, [33] and fan speed control [19].

2.1.4 Aging Effect

Apart from the variation of environmental conditions (e.g., voltage, temperature), chips also suffer from temporary or permanent changes in the structures of devices and interconnects over time, known as aging effects. These effects cause both short-term and long-term changes in multiple parameters of devices and circuits, leading to performance degradation and shortening the chip's lifetime. Moreover, high voltage and high temperature accelerate many aging effects, further exacerbating the problems. In the following, we elaborate on some of the main aging mechanisms.

- **Oxide defects**: The imperfection in the gate oxide lattice and the oxide-channel interface can contain defects which can capture and emit individual charges. Each defect can be characterized by its position, charge capture time (τ_c), and charge emission time (τ_e) [34, 35]. Depending on these properties, the charging and discharging of individual gate oxide defects cause a handful of effects (RTN, BTI, and HCI), leading to the change of multiple device parameters over time.

  Stochastic charging and discharging of oxide defects of short time constants (τ_c, τ_e) are manifested by *Random Telegraph Noise* (RTN), which is the phenomenon of high-frequency random V_{th} fluctuation [72]. RTN occurs when the system is in dynamic equilibrium, as shown in Fig. 2.6a [34]. It induces delay fluctuation of logic gates, which can reach 10% in a measurement of ring oscillator [43].

  In contrast, *Bias Temperature Instability* (BTI) is the directional V_{th} degradation and recovery over time, attributing to the charging and discharging of oxide defects of both short and long time constants in the nonequilibrium state, as shown in Fig. 2.6b. It can be further classified into Negative Bias Temperature Instability (NBTI) and Positive Bias Temperature Instability (PBTI) when occurring on PMOS and NMOS, respectively [16]. Typically, the long-term

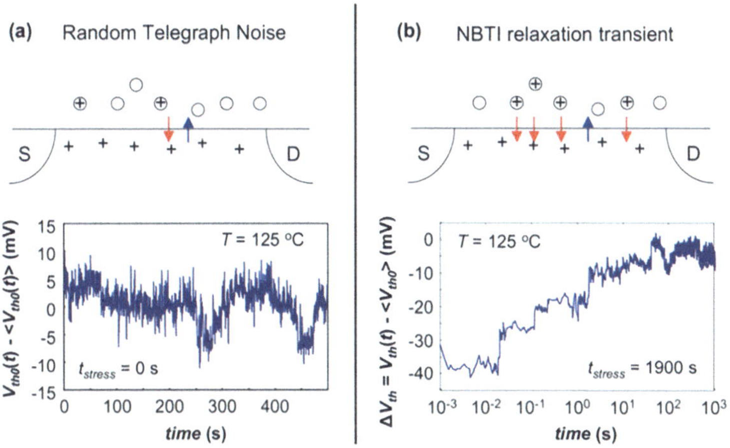

Fig. 2.6 The mechanisms of RTN and BTI. Replotted from [34]. (**a**) The system is in dynamic equilibrium, manifested by Random Telegraph Noise (RTN) in small devices. (**b**) With a gate voltage stress, the system is approaching the dynamic equilibrium with the gradual discharging of excess charged oxide defects, known as BTI

trend of BTI tends to gradually shift toward increased V_{th} and lowered carrier mobility. As a result, the chip's lifetime can be shortened due to BTI effects.

Hot Carrier Injection (HCI) is also caused by oxide charge trapping, but the mechanism is widely different. HCI originates from the high lateral electric field in the channel, normally due to high drain-source bias (V^{DS}). Carriers accelerate in the field and obtain high kinetic energy. These *hot carriers* can inject directly into the gate oxide or create high energy electrons and holes due to scattering, which in turn can also be injected into the oxide [15, 29, 31]. In consequence, changes accumulate in the gate oxide near the drain region, while the untrapped carriers contribute to gate leakage current and substrate leakage current. Similar to BTI, trapped changes caused by HCI result in increased V_{th} and reduced carrier mobility [46].

With the downscaling of the devices, the number of defects is reduced to only a handful. As a result, the stochastic properties become prominent, resulting in large variations in device performance and lifetime [24]. Figure 2.7 shows an example of the scaling effects of BTI. The V_{th} variability transforms from gradual changes as a combined effect of hundreds of effects, to a combination of individual discrete steps, each corresponding to a single charging/discharging event. The V_{th} variability also becomes dispersed among devices. Moreover, the effects of RTN, BTI, and HCI are highly dependent on voltage, temperature, and workload, which further complicates the estimation of the chip's performance degradation and lifetime.

- **Time-Dependent Dielectric Breakdown**: Time-Dependent Dielectric Breakdown (TDDB) is the phenomenon that damages accumulated in insulators induce

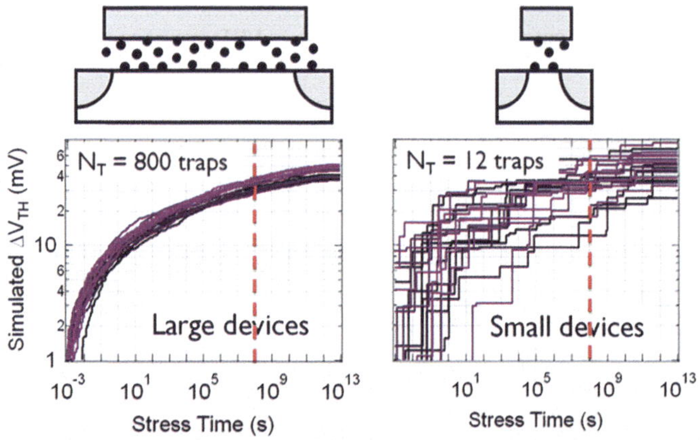

Fig. 2.7 The downscaling of devices reduces the number of defects to only a handful. As a result, the stochastic properties become prominent, resulting in large variations in device performance and lifetime. (Replotted from [24])

leakage current and eventually cause the loss of insulating property of the material [61]. It occurs in both the gate oxide of the devices and the low-k dielectrics of interconnects [4].

In the gate oxide, the injected charges can lead to trap-assisted leaking paths, increasing the gate leakage current steadily. This phenomenon is called Stress-Induced Leakage Current (SILC) [53]. Afterward, as trapped minority carriers are gradually built up in the damages of the gate oxide, conduction paths can be formed between the gate and the substrate when the vertical placement of the oxide changes is aligned. At this moment, the channel current is suddenly cut down, ending up with intermittent or permanent faults. This effect is called a hard breakdown, which poses a threat to the chip's lifetime. The gate oxide breakdown is influenced by supply voltage, device temperature, oxide thickness, and oxide materials [68, 76]. With the thinning down of gate oxide in advanced technologies, this effect is also exacerbated and presents as a serious reliability problem.

Similar mechanisms also occur in the dielectric of interconnects. Atoms of the conductors (e.g., copper) can infiltrate into the dielectric and create conductive paths between wires, leading to leakage current and thus increasing wire delay. Eventually, the hard breakdown occurs, and the insulating property is lost. This process can be accelerated by high temperature and high electric field. Some features in advanced technologies further aggravate the effect. First, the reduction of interconnect width and thickness increases the current density, while the smaller pitch increases the electric field. Second, the weaker breakdown strength of the porous, low-k dielectrics also enhances the effect [4].

- **Electromigration**: Electromigration (EM) is the phenomenon of undesired displacement of the conductor's atoms in the interconnects caused by electron flow [32, 63, 65]. As the electrons move through interconnects, they can collide with the atoms of the conductor. As a result, the atoms are displaced, leaving voids in their original positions and creating extrusions in the regions where these atoms drift to. Both the voids and the extrusions induce variation in resistance, which affects the chip's performance. If the displaced atoms further accumulate, the voids can cause open circuits, and the extrusions can cause short circuits between neighboring interconnects. In consequence, the chip could break down. With the shrinking of interconnects enabled by advanced lithography, the current density is increased, which enhances the EM effect. Therefore, EM becomes a main problem of the reliability of interconnects.

2.2 Architecture-Level Performance Variability

The ever-growing number of transistors has fueled the innovation and sophistication of the processor architecture, improving the throughput and versatility of processors. In a single processor core, the single-thread performance is enhanced by various microarchitectural features such as pipelining, out-of-order execution, speculative execution, and instruction-level parallelism. Multicore processors enable simultaneous executions of multiple threads in different cores, further boosting the throughput. Furthermore, system-on-chip (SoC) architecture integrates various functional units (e.g., GPU) in one chip, making processors more versatile [22]. Recently, the development of heterogeneous multicore architectures and domain-specific accelerators (DMAs) further improves the throughput and energy efficiency in specialized domains, such as artificial intelligence (AI) [51, 66]. Apart from the computation cores, memory hierarchy minimizes the average latency of data access, and interconnection networks facilitate data transfers among all units.

To illustrate the complexity of modern processor chips, Fig. 2.8 shows the block diagrams of ARM Zen 2 "Matisse" SoC architecture [62], a general-purpose processor designed for PC. It contains 16 processor cores, three levels of caches (either private or shared), interconnection networks, and several miscellaneous functional units (e.g., memory controller, IO controller).

With so many units and convoluted interactions, it is formidable to analyze the exact execution time of a workload. At the microarchitecture level, each unit has its own features of maximizing average throughput by exploiting dynamism, leading to variable latency [75]. At the macroarchitecture level, multiple units can contend for the same shared resources (e.g., caches, communication networks), so that one unit could delay the execution of another one [59, 75]. All of these architectural features contribute to performance variability, which is problematic for ensuring timing guarantees.

Branch prediction is a typical example of an architectural feature which causes performance variability. When a branch instruction is encountered in a processor core, the next instructions to be executed depend on the branch outcome, which would only be available when the branch instruction is resolved. Waiting for

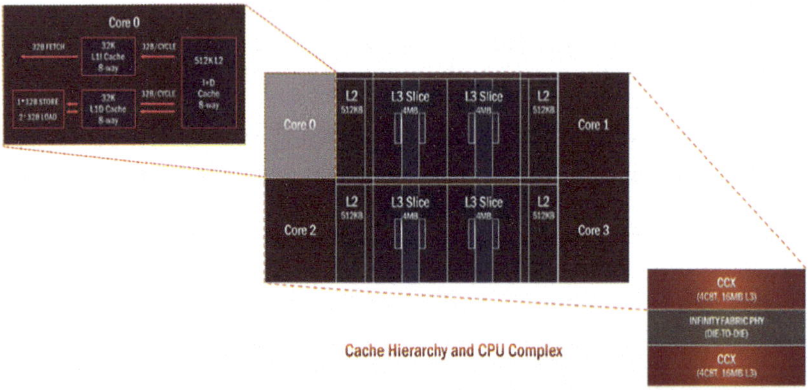

(a)

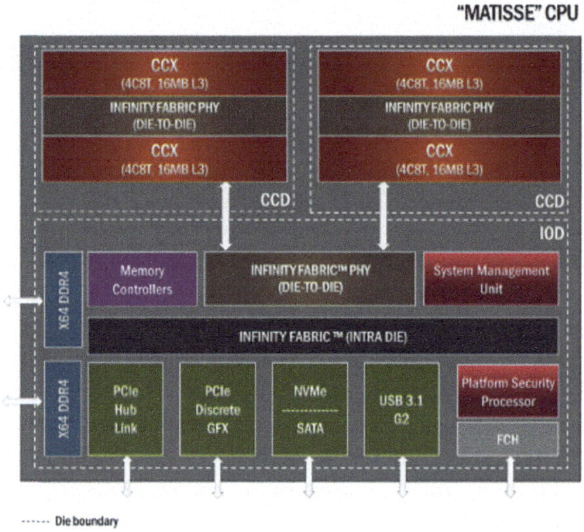

(b)

Fig. 2.8 Block diagrams of ARM Zen 2 "Matisse" SoC architecture. (Replotted from [62]). (**a**) The architecture of a "core complex" (CCX). (**b**) The architecture of the whole SoC, composed of four CCXs. Infinity fabric is a proprietary interconnection architecture which facilitates data transfers among linked components

the branch outcome would cause unfilled pipelines, resulting in several cycles of latency. Since branches are very common in programs, these latencies could accumulate to a significant throughput loss. Hence, predicting the branch outcome and speculatively executing the instructions of the predicted path can reduce these latencies. If the branch predictor is accurate, the latencies can be mostly removed, and the throughput can be substantially boosted.

To illustrate the performance variability generated by branch prediction, Fig. 2.9 shows a simple example of a program containing one branch. If the branch

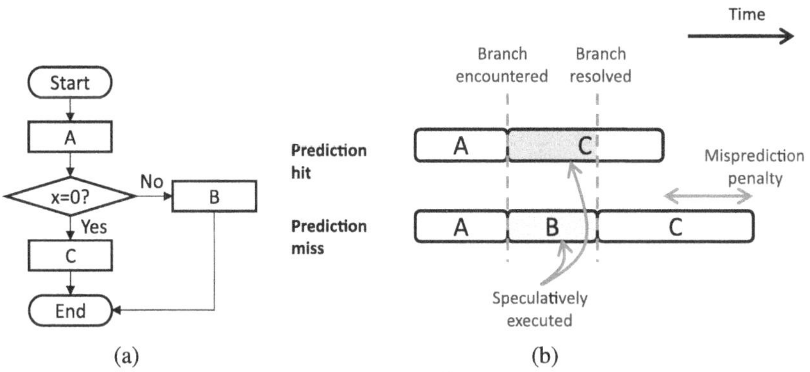

Fig. 2.9 Illustrative example of performance variability generated by branch prediction: (**a**) control flow of a task containing one branch and (**b**) time diagram of the processor in cases of either branch prediction hit or miss

prediction hits, the speculative execution outcomes become valid, and no time is wasted. On the contrary, if the branch prediction misses, the speculative execution outcomes have to be discarded. Thus, the time spent in the wrong branch directions is wasted, which is called the branch misprediction penalty. Evidently, whether branch prediction hits or misses will affect the execution time of the program. However, in order to improve prediction accuracy, the policy of branch predictors is complicated, often involving multiple strategies which can be dynamically switched depending on the application context and the processor states [44]. Therefore, it is difficult to completely characterize the behaviors of the branch predictors in advance, resulting in uncertainty in execution time.

2.3 Software-Level Performance Variability

Performance variability is manifested in the OS, middleware, and applications in many manners. First of all, since the OS is in charge of the allocation of the CPU time and hardware resources among processes, the execution time of an application will be highly dependent on other coexisting processes and the arbitration policy of the OS. In addition, the OS can interrupt the application with system administration tasks (e.g., system health monitoring), resulting in the increase and uncertainty in the execution time of the application [59].

During the execution of the application, different execution paths of the control flow can be taken according to the input data, leading to widely dispersed execution times. For example, the loop bound of a loop can be dependent on a certain input variable. Similar to the dynamism in the architecture, dynamic control flows are extensively exploited to improve performance and realize various functions, making the execution time difficult to analyze and predict.

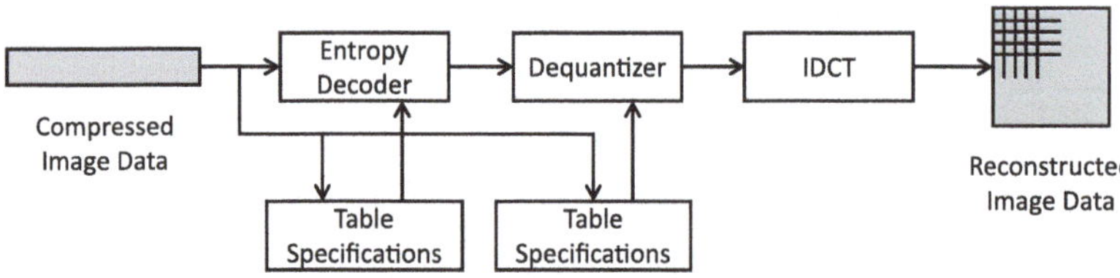

(a)

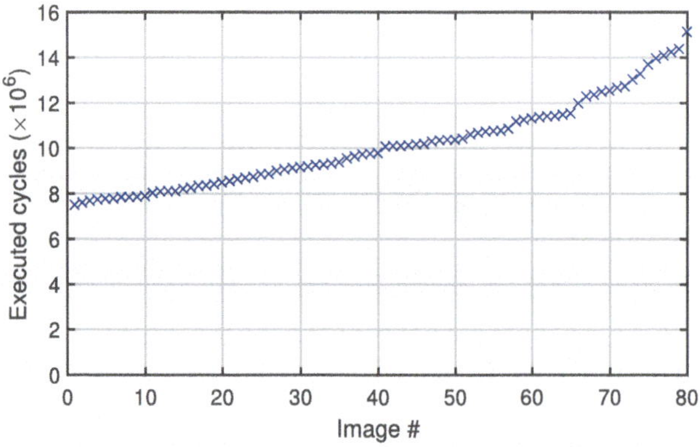

(b)

Fig. 2.10 Performance variability of a JPEG decoder application [20]: (**a**) block diagram of the JPEG decoder. Replotted from [28]. (**b**) Execution cycles of the JPEC decoder, benchmarked with different input images of the same size on a RISC-V ISA simulator, assuming $IPC = 1$ [52]

As a representative example, Fig. 2.10b shows the execution cycles of a JPEG decoder application [20], benchmarked with different input images of the same size on a RISC-V ISA simulator [52], assuming instruction per cycle (IPC) equals one. In this simulation, the highest execution time is approximately $2\times$ of the lowest one, exhibiting substantial performance variability. Within the program flow of JPEG decoding (shown in Fig. 2.10a [28]), the main source of performance variability comes from the entropy decoder block, in which Huffman decoding is performed based on the Huffman tables and the encoded bit string in the input data. Hence, the computation required for decoding is highly dependent on the input data.

2.4 Approaches to Tackling Variability

Many approaches have been developed to alleviate the undesired effects of variability. These approaches are instantiated on similar principles and tailored for different kinds of variability. In the following, we introduce some of the most important types of approaches to tackling variability.

2.4.1 Guardbanding

Guardbanding is the most straightforward method to deal with variability. In guardbanding, an upper bound of the variability among all possible conditions is calculated, which is called the *guardband* or *margin* [73]. In design-time, the chips are designed to function correctly within the specifications (i.e., design corners), which are defined as the nominal condition plus the guardband, so that the chip will not fail even in the circumstance of maximum variability. The guardband can be defined in different ways. Analytically, the guardband can be calculated from the worst-case scenario analysis or defined as certain quantiles (normally 3σ or 6σ) of the presumed statistical distribution of variability. Practically, it can also be determined according to the worst value from empirical observations or simply defined as an arbitrary value from industrial conventions, typically 10%.

The use of guardbanding is prevalent in every variability field. For example, the corner-based design paradigm requires the chips to work under the circumstance that all device parameters are deviated from their nominal values by 3σ in process variation [55]. To account for voltage variation, chips are designed to work even when the supply voltage is 10% deviated from its nominal level. For the BTI effect, a guardband of V_{th} is obtained assuming devices are under the DC stress at 125 °C over 3 years [36]. In tackling performance variability, WCET analysis is also a manifestation of guardbanding [2, 75].

Guardbanding is conventionally the main approach, if not the sole approach, to dealing with variability. However, it is no longer sufficient for the great magnitude and complexity of variability in modern chips. With the increasing amount of every variability source, the guardband can be greatly enlarged to 2× or more, significantly degrading the chip's quality or even making the chip's design infeasible. Figure 2.11 instantiates the failing of guardbanding in SRAM design, in which the guardbands for the worst case of each variability source sum up to an enormous voltage margin, which invalidates the scaling of V_{dd} in advanced technology [1].

The necessity for large guardbands is also questionable. In fact, guardbanding can lead to *over-margining*, which means the assumed worst-case scenario is too

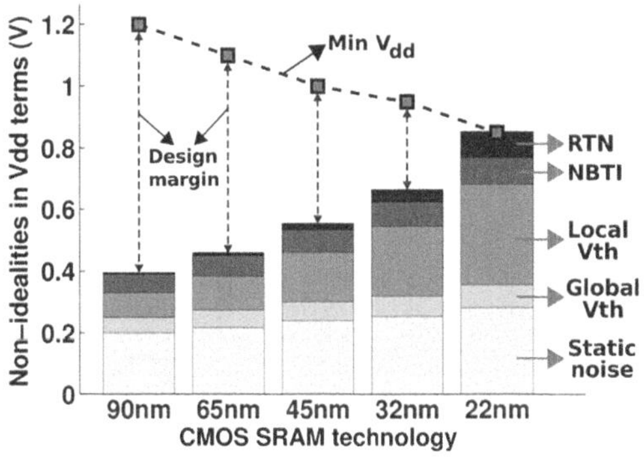

Fig. 2.11 Impact of each nonideality (quantified in V_{dd} terms) on SRAM supply voltage guardband under different CMOS technologies. The total guardband would reach the same level as the minimum V_{dd} with the technology downscaling, limiting the voltage scaling. (Replotted from [1])

conservative that it would never happen in the real world. Therefore, people have been working on more sophisticated and accurate variability models, striving to eliminate the pessimism in the guardbands as much as possible. The results are deemed to be limited since many variability mechanisms are dependent on the run-time environment and workload, which is agnostic at design-time.

2.4.2 Run-Time Mitigation

Since we can only know the actual amount of variability at run-time, it is sensible to deal with variability at run-time as well. Therefore, run-time mitigation arises from the idea to make the chips observe and mitigate the variability continuously [13], operating on similar principles of adaptive control systems.

Figure 2.12 shows the block diagram of a run-time mitigation system, in which monitors are distributed across widespread positions in the processor to monitor the amount of local variability in real time. The types of monitors are diverse, depending on the type of variability to observe. For example, voltage sensors and temperature sensors are deployed to directly detect on-chip voltage and temperature variations. In contrast, in situ critical path monitors and canary circuits are used to detect performance fluctuation of the circuit, which implies the variations in voltage, temperature, and aging [6, 9, 18, 41].

The real-time variability information is read by the run-time controller, which runs an adaptive control algorithm to determine the system's reaction to the variability. The algorithm makes decisions based on multiple factors, including the variability and the workload of the past, present, and predicted future. The controller is usually a small routine process in the OS or the firmware, while many modern processors are also equipped with a dedicated microcontroller to run the controller.

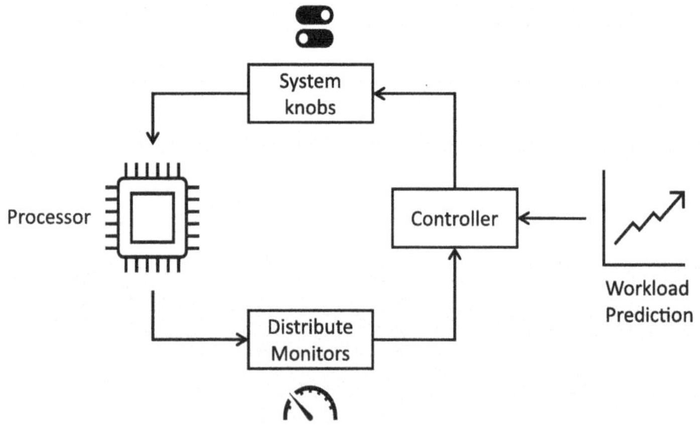

Fig. 2.12 Block diagram of a run-time mitigation system. (Replotted from [13])

Processors are equipped with system knobs, which can dynamically change the system parameters. For example, configurable clock generators and voltage converters allow the clock frequency and supply voltage to be altered at run-time. Based on the decision of the controller, the system knobs can be switched, by which system parameters are dynamically changed to mitigate variability.

On the principle of run-time mitigation, many variability management approaches have been developed and widely used to effectively deal with different kinds of variability. To prevent voltage droop, many voltage droop mitigation approaches have been developed and applied in modern integrated systems [7, 8, 21, 23, 25, 41, 57, 64, 69, 74]. As for temperature fluctuation, dynamic thermal management techniques are used to maintain chip temperature within operating region [11, 14, 40, 42, 54, 77]. In terms of lifetime, dynamic reliability management techniques are developed to mitigate long-term aging effects and prolong the lifetime of the chips [5, 26, 27, 37, 60]. To ensure timing guarantees, mitigation approaches for performance variability are also developed.

In terms of the effectiveness of meeting the requirements, the mitigation approaches can be categorized into three classes: the ones that cannot provide hard guarantees, the ones where at least an attempt is made to meet hard guarantees of real-time requirements, and the ones that can provide true timing guarantees. The first class is referred to as the best-effort (BE) approaches. Since normally these approaches are based on heuristics and these effects are only validated empirically, they cannot provide hard guarantees of their effectiveness. Most of the mitigation approaches proposed in the past fall into this class. The second class, such as WCET-based approaches, attempts to find the bounds of the variability and mitigate them. However, with oversimplified assumptions of real-time variability, they cannot ensure that the requirements are always met in practice. The last class, while being the most ideal, has not been realized in the past. However, we suppose workload-dependent approaches are the most promising option to fill this gap, which we will discuss in detail in the next chapter.

2.5 Summary

This chapter gives a comprehensive overview of the variability sources in each abstraction layer of integrated circuits. Firstly, the device- and circuit-level variability sources are enumerated, including process, voltage, and temperature variation, as well as several aging effects such as oxide defects, TDDB, and EM. These variations affect multiple device parameters (V_{th}, leakage current, etc.), circuit performance, and chip's lifetime. With the downscaling of the device dimensions, the magnitudes of these variations are mostly enlarged. Secondly, architecture-level performance variability originates from the architectural features of dynamic execution (e.g., branch prediction) and the contention of shared resources (e.g., caches). These features introduce uncertainty and unpredictability in execution time. With the increasing integration and complexity of processor architectures, the variability is exacerbated. Finally, software-level performance variability can be attributed to the resource allocation of the OS and the dynamic control flows of the applications, making the execution time difficult to analyze and predict.

To cope with variability, guardbanding is widely used to ensure the reliability of circuit design under the circumstance of the worst-case variability. However, large guardbands would degrade the quality of circuit design, and with the increasing variability, this problem is not acceptable anymore. Therefore, run-time mitigation solves the problem by allowing the chip to mitigate and adapt to the variability at run-time. Based on this principle, many approaches have been developed to effectively deal with different types of variability such as voltage, temperature, aging, and performance variability.

References

1. Aadithya, K.V., Demir, A., Venugopalan, S., Roychowdhury, J.: SAMURAI: an accurate method for modelling and simulating non-stationary Random Telegraph Noise in SRAMs. In: 2011 Design, Automation & Test in Europe, pp. 1–6 (2011). https://doi.org/10.1109/DATE.2011.5763183
2. Abella, J., Hernandez, C., Quiñones, E., Cazorla, F.J., Conmy, P.R., Azkarate-askasua, M., Perez, J., Mezzetti, E., Vardanega, T.: WCET analysis methods: pitfalls and challenges on their trustworthiness. In: 10th IEEE International Symposium on Industrial Embedded Systems (SIES), pp. 1–10 (2015). https://doi.org/10.1109/SIES.2015.7185039
3. Asenov, A., Brown, A., Davies, J., Kaya, S., Slavcheva, G.: Simulation of intrinsic parameter fluctuations in decananometer and nanometer-scale MOSFETs. IEEE Trans. Electron Devices **50**(9), 1837–1852 (2003). https://doi.org/10.1109/TED.2003.815862
4. Bekiaris, D., Papanikolaou, A., Papameletis, C., Soudris, D., Economakos, G., Pekmestzi, K.: A temperature-aware time-dependent dielectric breakdown analysis framework. In: van Leuken, R. Sicard, G. (eds.) Integrated Circuit and System Design. Power and Timing Modeling, Optimization, and Simulation, pp. 73–83. Springer, Berlin, Heidelberg (2011)
5. Bolchini, C., Carminati, M., Miele, A., Das, A., Kumar, A., Veeravalli, B.: Run-time mapping for reliable many-cores based on energy/performance trade-offs. In: 2013 IEEE International Symposium on Defect and Fault Tolerance in VLSI and Nanotechnology Systems (DFTS), pp. 58–64 (2013). https://doi.org/10.1109/DFT.2013.6653583
6. Bowman, K., Tschanz, J., Wilkerson, C., Lu, S.L., Karnik, T., De, V., Borkar, S.: Circuit techniques for dynamic variation tolerance. In: 2009 46th ACM/IEEE Design Automation Conference, pp. 4–7 (2009). https://doi.org/10.1145/1629911.1629915
7. Bowman, K.A., Raina, S., Bridges, J.T., Yingling, D.J., Nguyen, H.H., Appel, B.R., Kolla, Y.N., Jeong, J., Atallah, F.I., Hansquine, D.W.: A 16 nm all-digital auto-calibrating adaptive clock distribution for supply voltage droop tolerance across a wide operating range. IEEE J. Solid-State Circuits **51**(1), 8–17 (2016). https://doi.org/10.1109/JSSC.2015.2473655
8. Bowman, K.A., Tokunaga, C., Karnik, T., De, V.K., Tschanz, J.W.: A 22 nm all-digital dynamically adaptive clock distribution for supply voltage droop tolerance. IEEE J. Solid-State Circuits **48**(4), 907–916 (2013). https://doi.org/10.1109/JSSC.2013.2237972
9. Bowman, K.A., Tschanz, J.W., Lu, S.L.L., Aseron, P.A., Khellah, M.M., Raychowdhury, A., Geuskens, B.M., Tokunaga, C., Wilkerson, C.B., Karnik, T., De, V.K.: A 45 nm resilient microprocessor core for dynamic variation tolerance. IEEE J. Solid-State Circuits **46**(1), 194–208 (2011). https://doi.org/10.1109/JSSC.2010.2089657

10. Brooks, D., Martonosi, M.: Dynamic thermal management for high-performance microprocessors. In: Proceedings HPCA Seventh International Symposium on High-Performance Computer Architecture, pp. 171–182 (2001). https://doi.org/10.1109/HPCA.2001.903261
11. Brooks, D., Martonosi, M.: Dynamic thermal management for high-performance microprocessors. In: Proceedings HPCA Seventh International Symposium on High-Performance Computer Architecture, pp. 171–182 (2001). https://doi.org/10.1109/HPCA.2001.903261
12. Brown, A.R., Idris, N.M., Watling, J.R., Asenov, A.: Impact of metal gate granularity on threshold voltage variability: a full-scale three-dimensional statistical simulation study. IEEE Electron Device Lett. **31**(11), 1199–1201 (2010). https://doi.org/10.1109/LED.2010.2069080
13. Catthoor, F., Groeseneken, G.: Will chips of the future learn how to feel pain and cure themselves? IEEE Design Test **34**(5), 80–87 (2017). https://doi.org/10.1109/MDAT.2017.2730841
14. Chantem, T., Dick, R.P., Hu, X.S.: Temperature-aware scheduling and assignment for hard real-time applications on MPSoCs. In: 2008 Design, Automation and Test in Europe, pp. 288–293 (2008). https://doi.org/10.1109/DATE.2008.4484694
15. Chen, K.L., Saller, S., Groves, I., Scott, D.: Reliability effects on MOS transistors due to hot-carrier injection. IEEE J. Solid-State Circuits **20**(1), 306–313 (1985). https://doi.org/10.1109/JSSC.1985.1052307
16. Cho, M., Lee, J.D., Aoulaiche, M., Kaczer, B., Roussel, P., Kauerauf, T., Degraeve, R., Franco, J., Ragnarsson, L.Å., Groeseneken, G.: Insight into N/PBTI mechanisms in sub-1-nm-EOT devices. IEEE Trans. Electron Devices **59**(8), 2042–2048 (2012). https://doi.org/10.1109/TED.2012.2199496
17. Dadgour, H., Endo, K., De, V., Banerjee, K.: Modeling and analysis of grain-orientation effects in emerging metal-gate devices and implications for SRAM reliability. In: 2008 IEEE International Electron Devices Meeting, pp. 1–4 (2008). https://doi.org/10.1109/IEDM.2008.4796792
18. Deb, A., Vermeulen, B., van Dijk, L.: Overview of health monitoring techniques for reliability. In: Proceedings of the Workshop on Early Reliability Modeling for Aging and Variability in Silicon Systems (ERMAVSS 2016), pp. 30–33 (2016)
19. Dousti, M.J., Pedram, M.: Power-aware deployment and control of forced-convection and thermoelectric coolers. In: 2014 51st ACM/EDAC/IEEE Design Automation Conference (DAC), pp. 1–6 (2014). https://doi.org/10.1145/2593069.2593186
20. Falk, H., Altmeyer, S., Hellinckx, P., Lisper, B., Puffitsch, W., Rochange, C., Schoeberl, M., Sørensen, R.B., Wägemann, P., Wegener, S.: TACLeBench: A benchmark collection to support worst-case execution time research. In: Schoeberl, M. (ed.) 16th International Workshop on Worst-Case Execution Time Analysis (WCET 2016). OpenAccess Series in Informatics (OASIcs), vol. 55, pp. 2:1–2:10. Schloss Dagstuhl–Leibniz-Zentrum für Informatik, Dagstuhl (2016)
21. Floyd, M.S., Restle, P.J., Sperling, M.A., Owczarczyk, P., Fluhr, E.J., Friedrich, J., Muench, P., Diemoz, T., Chuang, P., Vezyrtzis, C.: Adaptive clocking in the POWER9^{TM} processor for voltage droop protection. In: 2017 IEEE International Solid-State Circuits Conference (ISSCC), pp. 444–445 (2017). https://doi.org/10.1109/ISSCC.2017.7870452
22. Furber, S.: Microprocessors: the engines of the digital age. Proc. Roy. Soc. A: Math. Phys. Eng. Sci. **473**, 20160893 (2017). https://doi.org/10.1098/rspa.2016.0893
23. Gonzalez, C., Floyd, M., Fluhr, E., Restle, P., Dreps, D., Sperling, M., Rao, R., Hogenmiller, D., Vezyrtis, C., Chuang, P., Lewis, D., Escobar, R., Ramadurai, V., Kruse, R., Pille, J., Nett, R., Owczarczyk, P., Friedrich, J., Paredes, J., Diemoz, T., Islam, S., Plass, D., Muench, P.: The 24-core POWER9 processor with adaptive clocking, 25-Gb/s accelerator links, and 16-Gb/s PCIe Gen4. IEEE J. Solid-State Circuits **53**(1), 91–101 (2018). https://doi.org/10.1109/JSSC.2017.2748623
24. Grasser, T., Kaczer, B., Goes, W., Reisinger, H., Aichinger, T., Hehenberger, P., Wagner, P.J., Schanovsky, F., Franco, J., Roussel, P., Nelhiebel, M.: Recent advances in understanding the bias temperature instability. In: 2010 International Electron Devices Meeting, pp. 4.4.1–4.4.4 (2010). https://doi.org/10.1109/IEDM.2010.5703295

25. Grenat, A., Pant, S., Rachala, R., Naffziger, S.: Adaptive clocking system for improved power efficiency in a 28nm x86-64 microprocessor. In: 2014 IEEE International Solid-State Circuits Conference Digest of Technical Papers (ISSCC), pp. 106–107 (2014). https://doi.org/10.1109/ISSCC.2014.6757358
26. Haghbayan, M.H., Miele, A., Rahmani, A.M., Liljeberg, P., Tenhunen, H.: Performance/reliability-aware resource management for many-cores in dark silicon era. IEEE Trans. Comput. **66**(9), 1599–1612 (2017). https://doi.org/10.1109/TC.2017.2691009
27. Hartman, A.S., Thomas, D.E.: Lifetime improvement through runtime wear-based task mapping. In: Proceedings of the Eighth IEEE/ACM/IFIP International Conference on Hardware/Software Codesign and System Synthesis, CODES+ISSS '12, p. 13–22. Association for Computing Machinery, New York (2012). https://doi.org/10.1145/2380445.2380455
28. Haskell, B., Howard, P., LeCun, Y., Puri, A., Ostermann, J., Civanlar, M., Rabiner, L., Bottou, L., Haffner, P.: Image and video coding-emerging standards and beyond. IEEE Trans. Circuits Syst. Video Technol. **8**(7), 814–837 (1998). https://doi.org/10.1109/76.735379
29. Heremans, P., Bellens, R., Groeseneken, G., Maes, H.: Consistent model for the hot-carrier degradation in n-channel and p-channel MOSFETs. IEEE Trans. Electron Devices **35**(12), 2194–2209 (1988). https://doi.org/10.1109/16.8794
30. Hu, C.: Low-voltage CMOS device scaling. In: Proceedings of IEEE International Solid-State Circuits Conference—ISSCC '94, pp. 86–87 (1994). https://doi.org/10.1109/ISSCC.1994.344716
31. Hu, C., Tam, S.C., Hsu, F.C., Ko, P.K., Chan, T.Y., Terrill, K.: Hot-electron-induced MOSFET degradation - model, monitor, and improvement. IEEE J. Solid-State Circuits **20**(1), 295–305 (1985). https://doi.org/10.1109/JSSC.1985.1052306
32. Hu, C.K., Rodbell, K.P., Sullivan, T.D., Lee, K.Y., Bouldin, D.P.: Electromigration and stress-induced voiding in fine Al and Al-alloy thin-film lines. IBM J. Res. Dev. **39**(4), 465–497 (1995). https://doi.org/10.1147/rd.394.0465
33. Iranfar, A., Terraneo, F., Csordas, G., Zapater, M., Fornaciari, W., Atienza, D.: Dynamic thermal management with proactive fan speed control through reinforcement learning. In: 2020 Design, Automation & Test in Europe Conference & Exhibition (DATE), pp. 418–423 (2020). https://doi.org/10.23919/DATE48585.2020.9116510
34. Kaczer, B., Grasser, T., Roussel, P.J., Franco, J., Degraeve, R., Ragnarsson, L.A., Simoen, E., Groeseneken, G., Reisinger, H.: Origin of NBTI variability in deeply scaled pFETs. In: 2010 IEEE International Reliability Physics Symposium, pp. 26–32 (2010). https://doi.org/10.1109/IRPS.2010.5488856
35. Kaczer, B., Mahato, S., Valduga de Almeida Camargo, V., Toledano-Luque, M., Roussel, P.J., Grasser, T., Catthoor, F., Dobrovolny, P., Zuber, P., Wirth, G., Groeseneken, G.: Atomistic approach to variability of bias-temperature instability in circuit simulations. In: 2011 International Reliability Physics Symposium, pp. XT.3.1–XT.3.5 (2011). https://doi.org/10.1109/IRPS.2011.5784604
36. Kang, K., Gangwal, S., Park, S.P., Kaushik, R.: NBTI induced performance degradation in logic and memory circuits: how effectively can we approach a reliability solution? In: 2008 Asia and South Pacific Design Automation Conference, pp. 726–731 (2008). https://doi.org/10.1109/ASPDAC.2008.4484047
37. Karl, E., Blaauw, D., Sylvester, D., Mudge, T.: Multi-mechanism reliability modeling and management in dynamic systems. IEEE Trans. Very Large Scale Integration Syst. **16**(4), 476–487 (2008). https://doi.org/10.1109/TVLSI.2007.915477
38. Kuhn, K.J.: Reducing variation in advanced logic technologies: approaches to process and design for manufacturability of nanoscale CMOS. In: 2007 IEEE International Electron Devices Meeting, pp. 471–474 (2007). https://doi.org/10.1109/IEDM.2007.4418976
39. Kuhn, K.J., Giles, M.D., Becher, D., Kolar, P., Kornfeld, A., Kotlyar, R., Ma, S.T., Maheshwari, A., Mudanai, S.: Process technology variation. IEEE Trans. Electron Devices **58**(8), 2197–2208 (2011). https://doi.org/10.1109/TED.2011.2121913

40. Lee, W., Patel, K., Pedram, M.: Dynamic thermal management for MPEG-2 decoding. In: ISLPED'06 Proceedings of the 2006 International Symposium on Low Power Electronics and Design, pp. 316–321 (2006). https://doi.org/10.1145/1165573.1165647
41. Lefurgy, C.R., Drake, A.J., Floyd, M.S., Allen-Ware, M.S., Brock, B., Tierno, J.A., Carter, J.B., Berry, R.W.: Active guardband management in Power7+ to save energy and maintain reliability. IEEE Micro **33**(4), 35–45 (2013). https://doi.org/10.1109/MM.2013.52
42. Liu, Z., Xu, T., Tan, S.X.D., Wang, H.: Dynamic thermal management for multi-core microprocessors considering transient thermal effects. In: 2013 18th Asia and South Pacific Design Automation Conference (ASP-DAC), pp. 473–478 (2013). https://doi.org/10.1109/ASPDAC.2013.6509641
43. Matsumoto, T., Kobayashi, K., Onodera, H.: Impact of random telegraph noise on CMOS logic circuit reliability. In: Proceedings of the IEEE 2014 Custom Integrated Circuits Conference, pp. 1–8 (2014). https://doi.org/10.1109/CICC.2014.6945997
44. McFarling, S.: Combining branch predictors. Tech. rep., Digital Western Research Laboratory (1993). URL https://www.hpl.hp.com/techreports/Compaq-DEC/WRL-TN-36.pdf
45. Muhtaroglu, A., Taylor, G., Rahal-Arabi, T.: On-die droop detector for analog sensing of power supply noise. IEEE J. Solid-State Circuits **39**(4), 651–660 (2004). https://doi.org/10.1109/JSSC.2004.825120
46. Mukherjee, S.: Architecture Design for Soft Errors. Morgan Kaufmann Publishers, San Francisco (2008)
47. Nam, H., Shin, C.: Study of high-k/metal-gate work-function variation using Rayleigh distribution. IEEE Electron Device Lett. **34**(4), 532–534 (2013). https://doi.org/10.1109/LED.2013.2247376
48. Nourani, M., Radhakrishnan, A.: Testing on-die process variation in nanometer VLSI. IEEE Design Test Comput. **23**(6), 438–451 (2006). https://doi.org/10.1109/MDT.2006.157
49. Ohmori, K., Matsuki, T., Ishikawa, D., Morooka, T., Aminaka, T., Sugita, Y., Chikyow, T., Shiraishi, K., Nara, Y., Yamada, K.: Impact of additional factors in threshold voltage variability of metal/high-k gate stacks and its reduction by controlling crystalline structure and grain size in the metal gates. In: 2008 IEEE International Electron Devices Meeting, pp. 1–4 (2008). https://doi.org/10.1109/IEDM.2008.4796707
50. Pant, S., Blaauw, D., Zolotov, V., Sundareswaran, S., Panda, R.: Vectorless analysis of supply noise induced delay variation. In: ICCAD-2003. International Conference on Computer Aided Design (IEEE Cat. No.03CH37486), pp. 184–191 (2003). https://doi.org/10.1109/ICCAD.2003.159688
51. Patterson, D.: 50 years of computer architecture: from the mainframe CPU to the domain-specific TPU and the open RISC-V instruction set. In: 2018 IEEE International Solid - State Circuits Conference - (ISSCC), pp. 27–31 (2018). https://doi.org/10.1109/ISSCC.2018.8310168
52. RISC-V International: Spike RISC-V ISA Simulator. https://github.com/riscv-software-src/riscv-isa-sim
53. Roussel, P., Degraeve, R., Van den Bosch, G., Kaczer, B., Groeseneken, G.: Accurate and robust noise-based trigger algorithm for soft breakdown detection in ultrathin gate dielectrics. IEEE Trans. Device Mater. Reliab. **1**(2), 120–127 (2001). https://doi.org/10.1109/7298.956706
54. Sabry, M.M., Atienza, D.: Temperature-aware design and management for 3D multi-core architectures. Found. Trends Electron. Design Autom. **8**(2), 117–197 (2014). https://doi.org/10.1561/1000000032
55. Saha, S.K.: Modeling process variability in scaled CMOS technology. IEEE Design Test Comput. **27**(2), 8–16 (2010). https://doi.org/10.1109/MDT.2010.50
56. Sakurai, T., Newton, A.: Alpha-power law MOSFET model and its applications to CMOS inverter delay and other formulas. IEEE J. Solid-State Circuits **25**(2), 584–594 (1990). https://doi.org/10.1109/4.52187

57. Shibahara, S., Takahashi, C., Fukuoka, K., Kitaji, Y., Irita, T., Hara, H., Shimazaki, Y., Matsushima, J.: A 16 nm FinFET heterogeneous nona-core SoC supporting ISO26262 ASIL B standard. IEEE J. Solid-State Circuits **52**(1), 77–88 (2017). https://doi.org/10.1109/JSSC.2016.2623682
58. Shin, C.: Variation-Aware Advanced CMOS Devices and SRAM. Springer, Dordrecht (2016)
59. Skinner, D., Kramer, W.: Understanding the causes of performance variability in HPC workloads. In: IEEE International. 2005 Proceedings of the IEEE Workload Characterization Symposium, 2005., pp. 137–149 (2005). https://doi.org/10.1109/IISWC.2005.1526010
60. Srinivasan, J., Adve, S., Bose, P., Rivers, J.: The case for lifetime reliability-aware microprocessors. In: Proceedings. 31st Annual International Symposium on Computer Architecture, 2004, pp. 276–287 (2004). https://doi.org/10.1109/ISCA.2004.1310781
61. Stathis, J.H.: Reliability limits for the gate insulator in CMOS technology. IBM J. Res. Dev. **46**(2.3), 265–286 (2002). https://doi.org/10.1147/rd.462.0265
62. Suggs, D., Subramony, M., Bouvier, D.: The AMD "Zen 2" processor. IEEE Micro **40**(2), 45–52 (2020). https://doi.org/10.1109/MM.2020.2974217
63. Sukharev, V., Najm, F.N.: Electromigration check: where the design and reliability methodologies meet. IEEE Trans. Device Mater. Reliab. **18**(4), 498–507 (2018). https://doi.org/10.1109/TDMR.2018.2874244
64. Takahashi, C., Shibahara, S., Fukuoka, K., Matsushima, J., Kitaji, Y., Shimazaki, Y., Hara, H., Irita, T.: A 16nm FinFET heterogeneous nona-core SoC complying with ISO26262 ASIL-B: achieving 10^{-7} random hardware failures per hour reliability. In: 2016 IEEE International Solid-State Circuits Conference (ISSCC), pp. 80–81 (2016). https://doi.org/10.1109/ISSCC.2016.7417916
65. Tao, J., Cheung, N., Hu, C.: Electromigration characteristics of copper interconnects. IEEE Electron Device Lett. **14**(5), 249–251 (1993). https://doi.org/10.1109/55.215183
66. Taylor, M.B.: Is dark silicon useful? harnessing the four horsemen of the coming dark silicon apocalypse. In: DAC Design Automation Conference 2012, pp. 1131–1136 (2012)
67. Tirumurti, C., Kundu, S., Sur-Kolay, S., Chang, Y.S.: A modeling approach for addressing power supply switching noise related failures of integrated circuits. In: Proceedings Design, Automation and Test in Europe Conference and Exhibition, vol. 2, pp. 1078–1083 (2004). https://doi.org/10.1109/DATE.2004.1269036
68. Torii, K., Kitajima, H., Arikado, T., Shiraishi, K., Miyazaki, S., Yamabe, K., Boero, M., Chikyow, T., Yamada, K.: Physical model of BTI, TDDB and SILC in HfO_2-based high-k gate dielectrics. In: IEDM Technical Digest. IEEE International Electron Devices Meeting, 2004, pp. 129–132 (2004). https://doi.org/10.1109/IEDM.2004.1419085
69. Tschanz, J., Kim, N.S., Dighe, S., Howard, J., Ruhl, G., Vangal, S., Narendra, S., Hoskote, Y., Wilson, H., Lam, C., Shuman, M., Tokunaga, C., Somasekhar, D., Tang, S., Finan, D., Karnik, T., Borkar, N., Kurd, N., De, V.: Adaptive frequency and biasing techniques for tolerance to dynamic temperature-voltage variations and aging. In: 2007 IEEE International Solid-State Circuits Conference. Digest of Technical Papers, pp. 292–604 (2007). https://doi.org/10.1109/ISSCC.2007.373409
70. Venkatachalam, V., Franz, M.: Power reduction techniques for microprocessor systems. ACM Comput. Surv. **37**(3), 195–237 (2005). https://doi.org/10.1145/1108956.1108957
71. Wang, X., Brown, A.R., Idris, N., Markov, S., Roy, G., Asenov, A.: Statistical threshold-voltage variability in scaled decananometer bulk HKMG MOSFETs: a full-scale 3-D simulation scaling study. IEEE Trans. Electron Devices **58**(8), 2293–2301 (2011). https://doi.org/10.1109/TED.2011.2149531
72. Weckx, P., Kaczer, B., Chen, C., Raghavan, P., Linten, D., Mocuta, A.: Relaxation of time-dependent NBTI variability and separation from RTN. In: 2017 IEEE International Reliability Physics Symposium (IRPS), pp. XT-9.1–XT-9.5 (2017). https://doi.org/10.1109/IRPS.2017.7936423
73. Weste, N., Harris, D.: CMOS VLSI Design: A Circuits and Systems Perspective, 4th edn. Addison-Wesley Publishing Company, USA (2010)

74. Wilcox, K., Cole, R., Fair III, H.R., Gillespie, K., Grenat, A., Henrion, C., Jotwani, R., Kosonocky, S., Munger, B., Naffziger, S., Orefice, R.S., Pant, S., Priore, D.A., Rachala, R., White, J.: Steamroller module and adaptive clocking system in 28 nm CMOS. IEEE J. Solid-State Circuits **50**(1), 24–34 (2015). https://doi.org/10.1109/JSSC.2014.2357428
75. Wilhelm, R., Engblom, J., Ermedahl, A., Holsti, N., Thesing, S., Whalley, D., Bernat, G., Ferdinand, C., Heckmann, R., Mitra, T., Mueller, F., Puaut, I., Puschner, P., Staschulat, J., Stenström, P.: The worst-case execution-time problem–overview of methods and survey of tools. ACM Trans. Embed. Comput. Syst. **7**(3) (2008). https://doi.org/10.1145/1347375.1347389
76. Yang, C.H., Chen, S.C., Tsai, Y.S., Lu, R., Lee, Y.H.: The physical explanation of TDDB power law lifetime model through oxygen vacancy trap investigations in HKMG NMOS FinFET devices. In: 2017 IEEE International Reliability Physics Symposium (IRPS), pp. 3C-4.1–3C-4.6 (2017). https://doi.org/10.1109/IRPS.2017.7936289
77. Yeo, I., Liu, C.C., Kim, E.J.: Predictive dynamic thermal management for multicore systems. In: 2008 45th ACM/IEEE Design Automation Conference, pp. 734–739 (2008)

Chapter 3
Mitigation Approaches for Performance Variability: State-of-the-Art

Abstract In this chapter, we elaborate on the state-of-the-art run-time mitigation approaches for tackling performance variability. In principle, these approaches aim to run workloads with just enough processor speed and hardware resources that the tasks could finish just in time. In this way, the minimum power and energy consumption can be reached to avoid thermal problems, reduce energy costs, and prolong the duration of battery-charged devices.

This chapter is structured as follows: First, we describe the general structure of mitigation approaches for performance variability in Sect. 3.1. Then, in Sect. 3.2 we discuss several typical system knobs used in performance variability mitigation. Finally, Sect. 3.3 discusses the heart of the mitigation approaches, the adaptive control algorithm for scheduling with and without timing guarantees.

3.1 Structure of Mitigation Approaches for Performance Variability

The mitigation approaches for performance variability follow the same principle of adaptive control systems, as discussed in Sect. 2.4.2, while the implementation is tailored for tackling performance variability. Figure 3.1 shows the block diagram of the performance variability mitigation system. The timer acts as the monitor of the system, measuring the execution time of each part of the application. At the beginning of the application, the run-time controller formulates an optimal schedule of knob configurations over time, according to the timing guarantee requirements and the predictions of future workload. During the execution, the run-time controller reads the timing information in the timer to calculate the amount of performance variability, by comparing the actual execution time with the predicted execution time. Depending on the amount of performance variability at run-time, the controller makes rolling decisions on adjusting the schedule and the knob configurations. By switching the knobs, the processor speed can be changed to offset the performance variability. In general, if the application progress lags behind the schedule, the controller can increase processor speed to catch up with the schedule. On the

J.-Y. Lin et al., *Workload Dependent Mitigation Approaches for Performance Variability*, https://doi.org/10.1007/978-3-031-85624-2_3

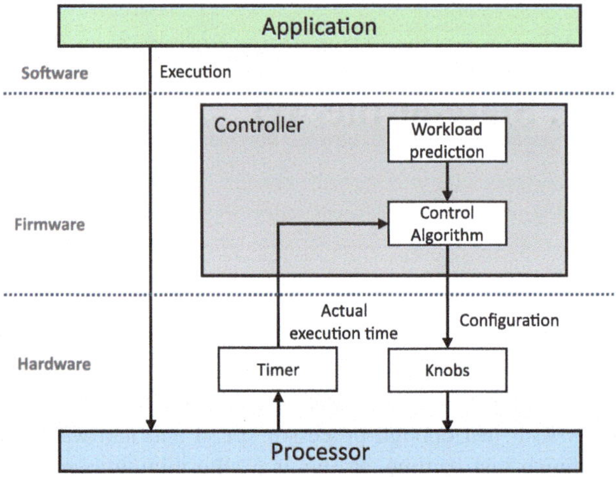

Fig. 3.1 Block diagram of the mitigation approach for performance variability

contrary, if the application progress is ahead of the schedule, the controller can slow down the processor to save energy.

3.2 System Knobs

The system knobs used in performance variability mitigation should be able to change the execution time of applications, either by adjusting the processor speed or by reallocating hardware resources to the applications. Depending on the application and the platform, the effect of system knobs can vary. For example, allocating different numbers of processor cores can substantially change the execution time of a parallelizable task, while a mostly sequential task is hardly affected.

These knobs are also different in switching time, which is the interval of changing the knob from one mode to another mode. Apparently, the switching time is essentially another source of performance variability which increases the execution time. When speeding up the processor, the new mode needs to take effect for a long enough time, so that the reduction of execution time can offset the original variability plus the switching time. Also, the switching time needs to be smaller than the timescale of the variability, in order not to significantly increase the total variability. Therefore, the timescale of the variability that a knob can handle is limited by its switching time.

The time granularity of the mitigation approaches is defined as the interval between two consecutive switchings of the knob. In general, the time granularity is much longer than the switching time, so the execution time overhead of knob switching is much smaller than the original workload. State-of-the-art approaches can operate down to the millisecond-level time granularity, limited by the switching time of dynamic voltage and frequency scaling (DVFS). To mitigate variability in more fine-grained time granularity, knobs with smaller switching time are required.

In the following, we discuss some of the most important knobs used in performance variability mitigation.

3.2.1 Dynamic Voltage and Frequency Scaling (DVFS)

Dynamic voltage and frequency scaling (DVFS) enables processor cores to dynamically change their supply voltage and clock frequency to vary their speed, power, and energy consumption [42]. The hardware setups of DVFS require a programmable clock generator, normally realized by a phase-locked loop (PLL), and a programmable voltage regulator, normally realized by a buck-type switching-mode DC-DC converter [24]. Since it is easy to use and provides a wide range of modes with substantial speed-power trade-offs, DVFS draws high popularity and is implemented in most modern processors. Table 3.1 shows a typical DVFS example in the Intel Core2 Duo processor, which contains six different voltage and frequency modes [24].

The speed-power variation of DVFS comes from the relation between dynamic power (P_{dyn}), supply voltage (V_{dd}), and clock frequency (f) of the circuit as below:

$$P_{\text{dyn}} = aCV_{\text{dd}}^2 f \tag{3.1}$$

where a is a coefficient reflecting the switching activity, and C is the equivalent capacitance of the circuit. Assuming the number of execution cycles (N_c) for a workload is constant regardless of the clock frequency, then the execution time (T) can be calculated as N_c/f. In this case, the dynamic power-induced energy consumption (E_{dyn}) will be equal to $P_{\text{dyn}} \times T$, resulting in the following relation:

$$E_{\text{dyn}} = N_c aCV_{\text{dd}}^2 \tag{3.2}$$

The relation between voltage and frequency of DVFS generally follows Eq. 2.3. Based on this relation and Eqs. 3.1 and 3.2, we can derive that P_{dyn} is quadratic to cubic with respect to V_{dd}, and E_{dyn} is quadratic with respect to V_{dd}. Since dynamic power is the dominant component of total power consumption in modern processor cores, a small shift in V_{dd} will bring substantial changes in clock frequency, total

Table 3.1 DVFS modes of the Intel Core2 Duo E6850 processor [24]

DVFS level	V_{dd} (V)	f (GHz)	DVFS level	V_{dd} (V)	f (GHz)
Level 1	1.30	3.074	Level 4	1.15	2.281
Level 2	1.25	2.852	Level 5	1.10	1.932
Level 3	1.20	2.588	Level 6	1.05	1.540

power, and energy. This property makes DVFS well suited for providing speed and power variations in processors.

Switching between DVFS modes requires time for the voltage and frequency to be stabilized at the new levels. When upscaling, the processor first configures the voltage regulator to increase the voltage to the new level, and then the PLL is configured to increase the frequency to the new level. When downscaling, the order is reversed in which the frequency is decreased first, and then the voltage is decreased. The switching time adds up to about 10–150 microseconds, depending on the DVFS modes before and after the transition [24]. While DVFS typically has the shortest switching time of all speed-energy-tradeoff knobs in the processors, it limits the time granularity of performance variability mitigation to the millisecond level, since the mitigation needs to operate in larger time granularity than the switching time.

Another drawback of DVFS resides in the incomplete freedom in selecting DVFS modes. In a multicore processor, multiple cores can share the same voltage domain, since the number of voltage domains in one chip is restricted by the number of package bumps. For correct operation, the shared voltage supply of the domain needs to cater to the core with the highest voltage demand, even if the other cores request lower voltage levels for power/energy saving [19]. On the other hand, the availability of high-voltage modes is also limited by the thermal constraint. It is normally disallowed for most of the cores to use high-voltage modes simultaneously. Otherwise, the massive collective power density can break down the chips [41]. These limitations result in the reduction and uncertainty in the performance power and energy benefits of DVFS.

3.2.2 Task Mapping and Task Migration

Heterogeneous multicore architectures enable great flexibility in the type and number of processor cores used by an application. By distributing workloads on multiple cores, parallelizable applications can be much accelerated, governed by Amdahl's law [2]. In addition, heterogeneous processor cores provide even wider options of the speed, power, and energy levels [3, 13]. Figure 3.2 shows an example of variability in execution time and energy consumption when running the same application with different core combinations [29]. With such a wide range of options to run every single task, the mapping of tasks to hardware resources has profound influences on the performance and energy efficiency of all tasks [38]. Furthermore, task mapping can be combined with DVFS to provide even more options to fine-tune the performance and energy efficiency of tasks.

Before the tasks start, the initial *task mapping* is solved to find the optimal task-to-core mapping in consideration of multiple objectives (performance, deadline constraints, energy, lifetime, etc.). During execution, *task migration* can be performed to reallocate the resources of the tasks, in order to react to run-time events, such as the entering of a new application [38]. To migrate a task, the controller first suspends the task and initiates a new process on the destination core. Then,

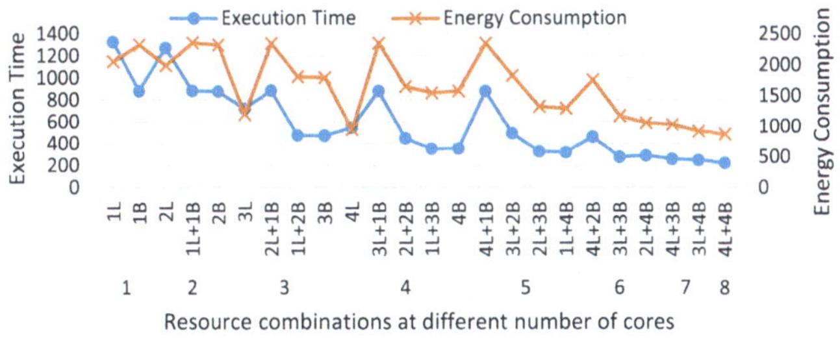

Fig. 3.2 Variability in execution time (seconds) and energy consumption (J) when running the Black-Scholes application (from PARSEC benchmark [7]) using different core combinations on ARM's big.LITTLE architecture which contains four big (B) and four LITTLE (L) cores. (Replotted from [29])

the whole task context (codes, data, stack, register file contents, processor states, etc.) is transferred from the source core to the destination cores. Finally, the task is continued on the destination core, and the data on the source core are purged [8, 23]. The switching time varies with the task context and the communication channels between the cores, typically up to tens of milliseconds [13].

The use of task mapping/migration also exhibits several important limitations. First, the millisecond-level switching time is much longer than DVFS, limiting the frequency of task migration. Also, considering the chip area costs, heterogeneous cores are normally only two or three types, which form very coarse-grained speed and power/energy levels. The acceleration enabled by multicore also depends on the exploitable parallelism in the application. For a mostly sequential application, multicore is ineffective.

Due to the above reasons, task mapping/migration is less flexible than DVFS, normally mitigating the performance variability of larger timescales. In general, task mapping/migration exploits the inherent performance and power/energy heterogeneity in macroarchitectures, so its coarse granularity is also attributed to the hefty sizes of macroarchitectures. To overcome these drawbacks, we suppose that it is possible to design fine-grained knobs by exploiting the heterogeneity in microarchitectures, such as heterogeneous datapaths (HDPs) which we will show later in our work.

3.2.3 *Gas Pedals*

Gas pedals, also known as *turbo modes* or *computational sprinting*, are short-lived high-performance modes which are available only for a limited duration due to system restrictions [28]. A typical instantiation of this concept is the ultrahigh voltage/frequency modes in DVFS, in which the high power dissipation could lead to thermal runaway if the modes run for an excessively long time. Gas pedals are implemented in some of the high-end commercial processors, such as Intel "Turbo Boost" technology [31, 32] and AMD "Precision Boost Overdrive" technology.

When the processor switches to the gas pedal mode, the massive power dissipation gradually raises the chip temperature. Therefore, the controller needs to turn off the gas pedal in time, so the chip would not fail due to excessively high temperature. Due to the reliability concern, gas pedals must be used with caution. Normally, the controller only turns on the gas pedal reactively when highly intensive workloads enter, which require extremely fast execution to meet the timing requirements. However, the timing guarantees can still be lost if the gas pedal is used too late. To ensure timing guarantees, the proactive way is to use the gas pedal as a resource which can be dynamically scheduled when intensive workloads are predicted in the upcoming future.

3.3 Adaptive Control Algorithms

In the run-time controller, the adaptive control algorithm is intended for finding the optimal way to switch the knobs which can ensure timing guarantees while minimizing energy consumption. The algorithm normally runs in the firmware or the OS as a background process. It is required to be lightweight so that its execution time is short enough to quickly respond to performance variability. Therefore, it is not practical to exhaust all possible scenarios to derive the optimal solution. Some heuristics need to be applied.

Different adaptive control algorithms can be considered as the trade-off between reliability and system costs. In the context of performance variability mitigation, the concerned reliability and cost metrics are the deadline hit rate and the energy consumption, respectively. The relation can be illustrated by the fact that in order to reach a high deadline hit rate, the algorithm needs to be more conservative so that the high-speed modes are used more often, increasing energy consumption.

Figure 3.3 depicts the comparison of reliability and cost among different types of algorithms. Best-effort (BE) approaches are algorithms without timing guarantees, which have low energy overhead but inferior deadline hit rate. On the other hand, WCET-based approaches belong to algorithms with full timing guarantees, which means they can achieve 100% hit rate whenever it is possible.

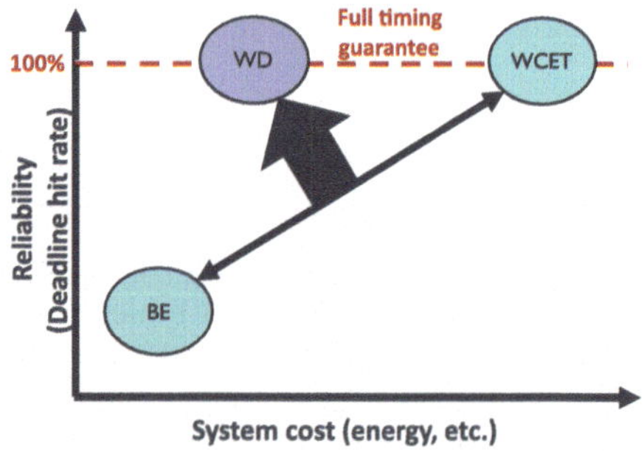

Fig. 3.3 Comparison of reliability (deadline hit rate) and cost (energy, etc.) among different types of adaptive control algorithms: best-effort (BE), worst-case execution time-based (WCET), and workload-dependent (WD)

However, WCET-based approaches lead to high energy overhead. To reduce the system costs, workload-dependent (WD) approaches aim to improve the execution time prediction based on run-time scenarios of the workload. In this way, energy consumption can be improved while the full timing guarantee is preserved. In the following subsections, we further discuss the state-of-the-art for each type of algorithm in detail.

3.3.1 Algorithms Without Timing Guarantees

Best-effort (BE) approaches are referred to as algorithms without timing guarantees, which aim to maintain stable performance of workloads with acceptable deadline miss rates while minimizing energy consumption. They cannot achieve the maximum deadline hit rate but can ensure guaranteed long-term throughput and near-optimal energy consumption. These approaches cannot reach the strict requirement of time-critical applications, so they are only feasible for general and time-sensitive applications.

The accuracy of workload prediction is crucial for the quality of BE approaches. If the actual execution time is not close to the prediction, the processor needs to dramatically speed up and down to compensate for the discrepancy in the workload progress generated by misprediction, resulting in extra energy consumption. Many heuristics are developed considering multiple factors including both design-time analysis and run-time observation in the past and present. In the following, we discuss several representative approaches:

- **PID-control**: The PID controller is widely used in various control system domains. On the same PID control principle, Rodopoulos et al. [30] proposed a PID-controlled DVFS algorithm which regulates the performance of a workload in the presence of performance variability. It is easy to implement since it operates based on only one variable the amount of performance variability at present. No prior knowledge of workload and no memory of past events are required. However, its effectiveness is also limited by its simple working principle, in that it can only reactively cope with performance variability after it occurs. Furthermore, its unawareness of the content and the execution history of the workload also degrades the quality of its decisions. These drawbacks limit its performance in preventing deadline misses and saving energy. To overcome these drawbacks, we have developed a PID mitigation approach working in parallel with an adaptive system scenario mechanism, which we will describe in Chap. 4.
- **History-based**: History-based approaches aim to predict the execution times of the upcoming jobs with the past jobs. For example, the predicted execution time of the next job can be the same as the last job [43], the weighted average of a couple of the most recent jobs [12], or the exponential moving average of the past jobs [25, 26]. The controller can also keep a record of the execution time distribution of the past jobs and then use the average [20] or a certain quantile

[45] as the prediction. Apparently, the prediction accuracy of these heuristics depends on the similarity of the future workload with the past workload. An abrupt change in the behaviors of workload would not be expected by these heuristics. Hence, these heuristics are more suitable for periodic, static, or slow-changing workloads, rather than highly dynamic workloads.

- **Machine learning**: With the rapid development of machine learning in recent years, approaches based on machine learning are also created to improve workload prediction accuracy and scheduling quality [18, 40]. The working principle of these approaches is to train a model which can translate the system states to the predicted processor performance or the optimal knob setting for the next interval. The training can be performed at design-time based on profiling data or at run-time based on online monitoring.

3.3.2 Algorithms with Attempts to Provide Timing Guarantees

To meet the ultimate timing requirements for time-critical applications, algorithms with attempts to provide full timing guarantees are developed, trying to achieve 100% hit rate whenever it is possible and reducing energy consumption at the same time. Timing-sensitive and general applications are also benefited from these algorithms since they ensure a guaranteed quality for these applications.

Figure 3.4 depicts the working principle of this type of algorithm, supposing that the controller can look into a buffer with the information of the upcoming n jobs. For each job in the buffer, an upper bound of execution time can be calculated to represent the worst-case scenario of each job. At the start of the workload, an optimal schedule is formulated, which can ensure no deadline misses even under the worst-case scenario that all jobs end up with their upper bound execution times. Then, Job 1 is executed under the modes assigned by the schedule. The actual execution time of Job 1 is usually smaller than the upper bound, so after Job 1 finishes, the actual progress of the workload is usually ahead of the schedule by an amount of time, which is called *slack*. The slack can be exploited by *slack reclaiming*, which means the schedule is refined by using the slack as an extra time resource to distribute it among the jobs in the buffer. Then, Job 2 is executed, which triggers another round of slack generation and reclaiming. With this repeated pattern, the controller keeps on rolling decisions until the workload ends.

To ensure timing guarantees, the key is to calculate the upper bound of execution time at run-time. The upper bound is used to derive the criteria in which the deadlines will never be missed. The upper bounds should be as tight as possible so that the discrepancy between the upper bound and the actual execution time is minimal, preventing the algorithm from being overconservative. In the calculation of upper bounds, WCET is used in most algorithms. However, WCET suffers from the problems of over-pessimism and oversimplification. Therefore, workload-dependent approaches have been proposed to improve the upper bound calculation. In the following, we discuss the two approaches in detail.

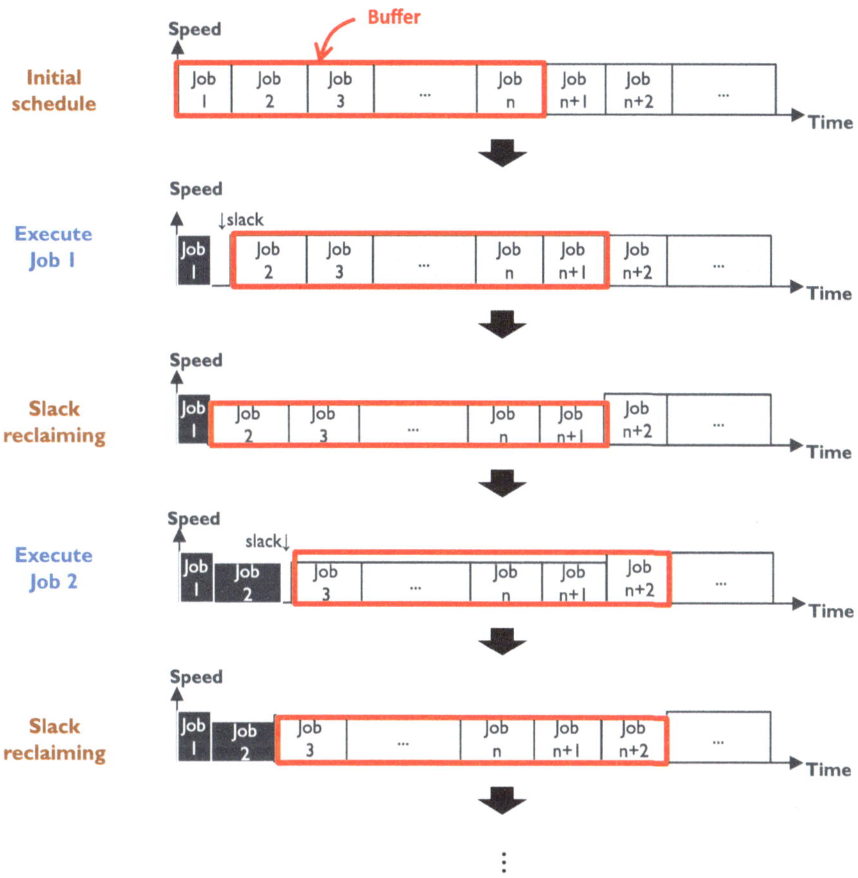

Fig. 3.4 Scheduling example of algorithms with timing guarantees

Worst-Case Execution Time (WCET)-Based Approaches

Worst-case execution time (WCET) analysis has been developed to obtain the maximum execution time of an application in all circumstances. Most approaches with timing guarantees rely on WCET to provide an upper bound of execution time. A typical WCET-based approach is a combination of real-time scheduling as the controlling algorithm, DVFS as the system knob, and WCET as the execution time bounding method.

Many scheduling algorithms have been developed based on the WCET [5]. In the finest granularity, several works [14, 21, 34, 46] have studied the optimal schedule of switching within a job which can minimize average energy consumption. However, these works usually overlook the time and energy overheads of switching, which makes these calculations impractical. For the scheduling of multiple tasks, the cycle-conserving algorithm of Pillai and Shin [27] marks out the basic principle of slack reclaiming in a fixed sequence of tasks. On this principle, real-time scheduling algorithms have been developed for both periodic [4, 16, 37, 48] and aperiodic [36] tasks. Scordino and Lipari [35] further proposed GRUB-PA, a generalized algorithm for aperiodic, sporadic, and periodic tasks, in which the processor speed is dynamically set based on the total utilization of the tasks. This algorithm has been implemented in the Linux kernel [33]. Recently, machine learning (ML) is also utilized to train a model which can dynamically choose the best one among multiple scheduling algorithms based on the system states [6, 10, 17, 47].

Despite the prevalence, WCET exhibits several important drawbacks. Firstly, WCET is normally much higher than the actual execution time. Measurements on real processors show that the observed maximum execution time of a common application can be 1.5–3 times the average execution time [15, 39]. On top of that, WCET estimates are reported to have overestimation by up to 50% compared to the actual maximum execution time [44]. Therefore, in the scheduling, assuming the WCET for all tasks leads to an exaggeration of the actual execution time, resulting in over-conservative decisions and large energy overheads.

Secondly, the complexity of modern architectures and software has been challenging for WCET analysis. There are several methods to perform WCET analysis. The static method analyzes the control flow of the application to find the longest execution path. Then, WCET is obtained by summing up the execution times of all steps along this path [1]. Though this method is the only way to derive a certified bound of execution time, the complication of modern architectures and applications makes the execution time formidable to analyze [44]. On the other hand, measurement-based and probabilistic methods obtain the WCET by estimating the maximum or the tailing part of the execution time distribution based on sampling data [1]. However, these methods cannot guarantee that the WCET is never underestimated [44], so they do not meet the utmost standard for time-critical applications.

Workload-Dependent Approaches

Workload-dependent approaches operate on the principle of using design-time information to improve the quality of run-time scheduling. Based on the reference of design-time information, the controller can dynamically eliminate unnecessary pessimism, reducing energy overheads while the timing guarantees are still ensured. Therefore, we suppose that workload-dependent approaches are key to realizing an algorithm with full timing guarantees and low system costs.

The amount and the types of information brought from design-time to run-time are crucial in the design of workload-dependent approaches. Bringing more information would improve the quality of scheduling, but it increases memory storage and complicates the calculation, inducing more execution time and energy overheads. Therefore, a workload-dependent approach should strike a balance between the scheduling quality and the run-time overheads.

Dynamic scenarios (DSs), a generic concept of systematically handling the dynamic nature of modern systems with jointly design-time/run-time efforts, has been adopted to solve this problem [22]. In this concept, similar system behaviors are grouped into scenarios at design-time. Then, at run-time, by identifying the scenarios to which the current and upcoming run-time situations belong, the system behaviors and costs can be more accurately calculated and predicted [9, 11]. Based on the principle of dynamic scenarios, cost-effective approaches can be developed for workload-dependent run-time prediction and calculation of execution time and energy consumption, which can dynamically trim the upper bound of execution time according to run-time scenarios. The dynamic scenario methodology serves as a cost-effective low-error way to transfer system cost information from design-time to run-time. Therefore, in our work, these principles are instantiated to develop the workload-dependent mitigation approach for performance variability.

Dynamic scenarios are developed from its predecessors, *system scenarios* and *adaptive scenarios*. System scenarios define the basic principles of design-time scenario clustering and run-time scenario identification. Adaptive scenarios extend the concept by allowing run-time scenario re-clustering based on past workload. Finally, dynamic scenario completes the conceptual framework by scenario prediction and identification of future workload. In the rest of the book, the concepts of these scenarios will be elaborated in detail. In Chap. 4, we will first introduce the concepts of system scenarios and adaptive scenarios, which are the precedents of dynamic scenarios. Then in Chap. 5, we will further elaborate on the concept of dynamic scenarios.

3.4 Summary

In this chapter, mitigation approaches for performance variability in state-of-the-art are reviewed, including the overall structure and the components. The general structure of the approaches is introduced, which is an adaptive control system composed of a timer, a run-time controller, and speed-varying knobs.

The most popular knobs are DVFS and task mapping/migration. DVFS varies the supply voltage and clock frequency at run-time to realize the speed-energy trade-off. It is easy to use and offers wide speed and energy ranges. However, its drawbacks lie in significant switching times (10–150 μs) and incomplete freedom in selecting DVFS modes. Task mapping/migration can change the number and the types of cores allocated to a workload. However, it has long switching times (up to tens of milliseconds), and its effectiveness is highly dependent on the exploitable parallelism in the application.

The adaptive control algorithms in the run-time controller can be classified into two groups, depending on whether they can ensure timing guarantees. Best-effort approaches rely on heuristics to minimize energy consumption to regulate the long-term throughput, but they cannot ensure timing guarantees. On the other hand, WCET-based approaches ensure timing guarantees by using WCET as the upper bound of execution time in scheduling. However, the pessimism in WCET makes the approaches overconservative and induces large energy overheads. Workload-dependent approaches are promising to solve this problem by bringing design-time information to help in run-time decisions. In this way, the upper bounds can be dynamically trimmed down to reduce pessimism and save energy.

References

1. Abella, J., Hernandez, C., Quiñones, E., Cazorla, F.J., Conmy, P.R., Azkarate-askasua, M., Perez, J., Mezzetti, E., Vardanega, T.: WCET analysis methods: Pitfalls and challenges on their trustworthiness. In: 10th IEEE International Symposium on Industrial Embedded Systems (SIES), pp. 1–10 (2015). https://doi.org/10.1109/SIES.2015.7185039
2. Amdahl, G.M.: Validity of the single processor approach to achieving large scale computing capabilities. In: Proceedings of the April 18–20, 1967, Spring Joint Computer Conference, AFIPS '67 (Spring), p. 483–485. Association for Computing Machinery, New York (1967). https://doi.org/10.1145/1465482.1465560
3. ARM: big.LITTLE technology: The future of mobile. Tech. rep., ARM white paper (2013). https://armkeil.blob.core.windows.net/developer/Files/pdf/white-paper/big-little-technology-the-future-of-mobile.pdf
4. Aydin, H., Melhem, R., Mosse, D., Mejia-Alvarez, P.: Dynamic and aggressive scheduling techniques for power-aware real-time systems. In: Proceedings 22nd IEEE Real-Time Systems Symposium (RTSS 2001), pp. 95–105 (2001). https://doi.org/10.1109/REAL.2001.990600
5. Bambagini, M., Marinoni, M., Aydin, H., Buttazzo, G.: Energy-aware scheduling for real-time systems: A survey. ACM Trans. Embed. Comput. Syst. **15**(1) (2016). https://doi.org/10.1145/2808231
6. Bhatti, M., Belleudy, C., Auguin, M.: Hybrid power management in real time embedded systems: an interplay of DVFS and DPM techniques. Real-Time Syst. **47**, 143–162 (2011). https://doi.org/10.1007/s11241-011-9116-y
7. Bienia, C., Li, K.: PARSEC 2.0: a new benchmark suite for chip-multiprocessors. In: Proceedings of the 5th Annual Workshop on Modeling, Benchmarking and Simulation (2009)
8. Briao, E.W., Barcelos, D., Wagner, F.R.: Dynamic task allocation strategies in MPSoC for soft real-time applications. In: 2008 Design, Automation and Test in Europe, pp. 1386–1389 (2008). https://doi.org/10.1109/DATE.2008.4484934
9. Catthoor, F., Basten, T., Zompakis, N., Geilen, M., Kjeldsberg, P.G.: System-Scenario-based Design Principles and Applications. Springer, Dordrecht (2020)
10. Dhiman, G., Rosing, T.S.: System-level power management using online learning. IEEE Trans. Comput.-Aided Design Integrated Circuits Syst. **28**(5), 676–689 (2009). https://doi.org/10.1109/TCAD.2009.2015740
11. Gheorghita, S.V., Palkovic, M., Hamers, J., Vandecappelle, A., Mamagkakis, S., Basten, T., Eeckhout, L., Corporaal, H., Catthoor, F., Vandeputte, F., Bosschere, K.D.: System-scenario-based design of dynamic embedded systems. ACM Trans. Design Autom. Electron. Syst. **14**(1) (2009). https://doi.org/10.1145/1455229.1455232
12. Govil, K., Chan, E., Wasserman, H.: Comparing algorithm for dynamic speed-setting of a low-power CPU. In: Proceedings of the 1st Annual International Conference on Mobile Computing and Networking, MobiCom '95, p. 13–25. Association for Computing Machinery, New York (1995). https://doi.org/10.1145/215530.215546
13. Greenhalgh, P.: big.LITTLE processing with ARM Cortex-A15 & Cortex-A7. Tech. rep., ARM white paper (2011)
14. Gruian, F.: Hard real-time scheduling for low-energy using stochastic data and DVS processors. In: ISLPED'01: Proceedings of the 2001 International Symposium on Low Power Electronics and Designd, pp. 46–51 (2001). https://doi.org/10.1109/LPE.2001.945370

15. Hapka, R., Christmann, A., Ernst, R.: Controlling high-performance platform uncertainties with timing diversity. In: 2022 IEEE 28th International Conference on Embedded and Real-Time Computing Systems and Applications (RTCSA), pp. 212–219 (2022). https://doi.org/10.1109/RTCSA55878.2022.00029
16. Im, C., Kim, H., Ha, S.: Dynamic voltage scheduling technique for low-power multimedia applications using buffers. In: ISLPED'01: Proceedings of the 2001 International Symposium on Low Power Electronics and Design, pp. 34–39 (2001). https://doi.org/10.1109/LPE.2001.945368
17. Islam, F.M.M.u., Lin, M.: Hybrid DVFS scheduling for real-time systems based on reinforcement learning. IEEE Syst. J. **11**(2), 931–940 (2017). https://doi.org/10.1109/JSYST.2015.2446205
18. Jung, H., Pedram, M.: Supervised learning based power management for multicore processors. IEEE Trans. Comput.-Aided Design Integrated Circuits Syst. **29**(9), 1395–1408 (2010). https://doi.org/10.1109/TCAD.2010.2059270
19. Kim, S.T., Shih, Y.C., Mazumdar, K., Jain, R., Ryan, J.F., Tokunaga, C., Augustine, C., Kulkarni, J.P., Ravichandran, K., Tschanz, J.W., Khellah, M.M., De, V.: Enabling wide autonomous DVFS in a 22 nm graphics execution core using a digitally controlled fully integrated voltage regulator. IEEE J. Solid-State Circuits **51**(1), 18–30 (2016). https://doi.org/10.1109/JSSC.2015.2457920
20. Kumar, P., Srivastava, M.: Predictive strategies for low-power RTOS scheduling. In: Proceedings 2000 International Conference on Computer Design, pp. 343–348 (2000). https://doi.org/10.1109/ICCD.2000.878306
21. Lee, S., Sakurai, T.: Run-time voltage hopping for low-power real-time systems. In: Proceedings 37th Design Automation Conference, pp. 806–809 (2000). https://doi.org/10.1145/337292.337785
22. Munaga, S., Catthoor, F.: Systematic design principles for cost-effective hard constraint management in dynamic nonlinear systems. Int. J. Adapt. Resilient Auton. Syst. **2**(1), 18–45 (2011). https://doi.org/10.4018/jaras.2011010102
23. Nollet, V., Marescaux, T., Avasare, P., Verkest, D., Mignolet, J.Y.: Centralized run-time resource management in a network-on-chip containing reconfigurable hardware tiles. In: Design, Automation and Test in Europe, vol. 1, pp. 234–239 (2005). https://doi.org/10.1109/DATE.2005.91
24. Park, S., Park, J., Shin, D., Wang, Y., Xie, Q., Pedram, M., Chang, N.: Accurate modeling of the delay and energy overhead of dynamic voltage and frequency scaling in modern microprocessors. IEEE Trans. Comput.-Aided Design Integrated Circuits Syst. **32**(5), 695–708 (2013). https://doi.org/10.1109/TCAD.2012.2235126
25. Pering, T., Burd, T., Brodersen, R.: The simulation and evaluation of dynamic voltage scaling algorithms. In: Proceedings. 1998 International Symposium on Low Power Electronics and Design, pp. 76–81 (1998). https://doi.org/10.1145/280756.280790
26. Pering, T., Burd, T., Brodersen, R.: Voltage scheduling in the IpARM microprocessor system. In: ISLPED'00: Proceedings of the 2000 International Symposium on Low Power Electronics and Design, pp. 96–101 (2000). https://doi.org/10.1145/344166.344530
27. Pillai, P., Shin, K.G.: Real-time dynamic voltage scaling for low-power embedded operating systems. In: Proceedings of the Eighteenth ACM Symposium on Operating Systems Principles, SOSP '01, p. 89–102. Association for Computing Machinery, New York (2001). https://doi.org/10.1145/502034.502044
28. Raghavan, A., Luo, Y., Chandawalla, A., Papaefthymiou, M., Pipe, K.P., Wenisch, T.F., Martin, M.M.K.: Computational sprinting. In: IEEE International Symposium on High-Performance Comp Architecture, pp. 1–12 (2012). https://doi.org/10.1109/HPCA.2012.6169031
29. Reddy, B.K., Singh, A.K., Biswas, D., Merrett, G.V., Al-Hashimi, B.M.: Inter-cluster thread-to-core mapping and DVFS on heterogeneous multi-cores. IEEE Trans. Multi-Scale Comput. Syst. **4**(3), 369–382 (2018). https://doi.org/10.1109/TMSCS.2017.2755619

30. Rodopoulos, D., Catthoor, F., Soudris, D.: Tackling performance variability due to RAS mechanisms with PID-controlled DVFS. IEEE Comput. Archit. Lett. **14**(2), 156–159 (2015). https://doi.org/10.1109/LCA.2014.2385713
31. Rotem, E., Naveh, A., Ananthakrishnan, A., Weissmann, E., Rajwan, D.: Power-management architecture of the Intel microarchitecture code-named Sandy Bridge. IEEE Micro **32**(2), 20–27 (2012). https://doi.org/10.1109/MM.2012.12
32. Rotem, E., Naveh, A., Rajwan, D., Ananthakrishnan, A., Weissmann, E.: Power management architecture of the 2nd generation Intel®core microarchitecture, formerly codenamed sandy bridge. In: 2011 IEEE Hot Chips 23 Symposium (HCS), pp. 1–33 (2011). https://doi.org/10.1109/HOTCHIPS.2011.7477510
33. Scordino, C., Abeni, L., Lelli, J.: Energy-aware real-time scheduling in the Linux kernel. In: Proceedings of the 33rd Annual ACM Symposium on Applied Computing, SAC '18, p. 601–608. Association for Computing Machinery, New York (2018). https://doi.org/10.1145/3167132.3167198
34. Scordino, C., Bini, E., et al.: Optimal speed assignment for probabilistic execution times. In: 2nd Workshop on Power-Aware Real-Time Computing (PARC'05). Citeseer (2005)
35. Scordino, C., Lipari, G.: A resource reservation algorithm for power-aware scheduling of periodic and aperiodic real-time tasks. IEEE Trans. Comput. **55**(12), 1509–1522 (2006). https://doi.org/10.1109/TC.2006.190
36. Shin, D., Kim, J.: Dynamic voltage scaling of periodic and aperiodic tasks in priority-driven systems. In: ASP-DAC 2004: Asia and South Pacific Design Automation Conference 2004 (IEEE Cat. No.04EX753), pp. 653–658 (2004). https://doi.org/10.1109/ASPDAC.2004.1337673
37. Shin, Y., Choi, K.: Power conscious fixed priority scheduling for hard real-time systems. In: Proceedings 1999 Design Automation Conference, pp. 134–139 (1999). https://doi.org/10.1109/DAC.1999.781298
38. Singh, A.K., Shafique, M., Kumar, A., Henkel, J.: Mapping on multi/many-core systems: Survey of current and emerging trends. In: 2013 50th ACM/EDAC/IEEE Design Automation Conference (DAC), pp. 1–10 (2013). https://doi.org/10.1145/2463209.2488734
39. Skinner, D., Kramer, W.: Understanding the causes of performance variability in HPC workloads. In: IEEE International. 2005 Proceedings of the IEEE Workload Characterization Symposium, 2005, pp. 137–149 (2005). https://doi.org/10.1109/IISWC.2005.1526010
40. Tarsa, S.J., Kumar, A.P., Kung, H.T.: Workload prediction for adaptive power scaling using deep learning. In: 2014 IEEE International Conference on IC Design & Technology, pp. 1–5 (2014). https://doi.org/10.1109/ICICDT.2014.6838580
41. Taylor, M.B.: Is dark silicon useful? Harnessing the four horsemen of the coming dark silicon apocalypse. In: DAC Design Automation Conference 2012, pp. 1131–1136 (2012)
42. Venkatachalam, V., Franz, M.: Power reduction techniques for microprocessor systems. ACM Comput. Surv. **37**(3), 195–237 (2005). https://doi.org/10.1145/1108956.1108957
43. Weiser, M., Welch, B., Demers, A., Shenker, S.: Scheduling for reduced CPU energy. In: Proceedings of the 1st USENIX Conference on Operating Systems Design and Implementation, OSDI '94. USENIX Association, Berkeley (1994)
44. Wilhelm, R., Engblom, J., Ermedahl, A., Holsti, N., Thesing, S., Whalley, D., Bernat, G., Ferdinand, C., Heckmann, R., Mitra, T., Mueller, F., Puaut, I., Puschner, P., Staschulat, J., Stenström, P.: The worst-case execution-time problem–overview of methods and survey of tools. ACM Trans. Embed. Comput. Syst. **7**(3) (2008). https://doi.org/10.1145/1347375.1347389
45. Yuan, W., Nahrstedt, K.: Energy-efficient soft real-time CPU scheduling for mobile multimedia systems. In: Proceedings of the Nineteenth ACM Symposium on Operating Systems Principles, SOSP '03, p. 149–163. Association for Computing Machinery, New York (2003). https://doi.org/10.1145/945445.945460

46. Yuan, W., Nahrstedt, K.: Energy-efficient soft real-time CPU scheduling for mobile multimedia systems. In: Proceedings of the Nineteenth ACM Symposium on Operating Systems Principles, SOSP '03, pp. 149–163. Association for Computing Machinery, New York (2003). https://doi.org/10.1145/945445.945460
47. Zhang, Q., Lin, M., Yang, L.T., Chen, Z., Li, P.: Energy-efficient scheduling for real-time systems based on deep Q-learning model. IEEE Trans. Sustain. Comput. **4**(1), 132–141 (2019). https://doi.org/10.1109/TSUSC.2017.2743704
48. Zhu, Y., Mueller, F.: Feedback EDF scheduling exploiting dynamic voltage scaling. In: Proceedings. RTAS 2004. 10th IEEE Real-Time and Embedded Technology and Applications Symposium, 2004, pp. 84–93 (2004). https://doi.org/10.1109/RTTAS.2004.1317252

Chapter 4
Mitigating Performance Variability Using System and Adaptive Scenarios

Abstract In this chapter, we will present the concept of system and adaptive scenarios, which are the predecessors of dynamic scenarios. Also, we will deploy a PID scheme on a fine workload granularity, applying Dynamic Voltage and Frequency Scaling (DVFS) actuations per basic workload chunks named as Thread Nodes (TNs); we believe that a TN-level mitigation of performance variation permits an accurate and cost efficient mitigation of performance variability.

The structure of this chapter is the following: In Sect. 4.1 we will first define the related terminology and discuss the foundations of the system scenario concept that groups TNs with similar cost perspectives. In Sect. 4.2, we will apply a TN analysis and a system scenario clustering on our spectrum sensing application while presenting related simulation results. The PID controller working in parallel with an *adaptive system scenario* mechanism, which allows run-time re-clustering decisions of the scenario hierarchy, will be studied later in Sect. 4.3; simulation results of Sect. 4.4 will show the benefit of the latter approach. Finally, Sect. 4.5 will conclude the chapter.

4.1 System Scenarios: General Concepts and Related Terminology

To elaborate on the system scenario methodology, we first need to discuss some basic concepts. For a more verbose definition of the following terms and extensive coverage, we refer readers to the previous related works [4, 5]

4.1.1 General Concepts

We can divide system workload into a number of thread nodes; thread node categorization constitutes a finer granularity compared to a task-level one. In short:

J.-Y. Lin et al., *Workload Dependent Mitigation Approaches for Performance Variability*, https://doi.org/10.1007/978-3-031-85624-2_4

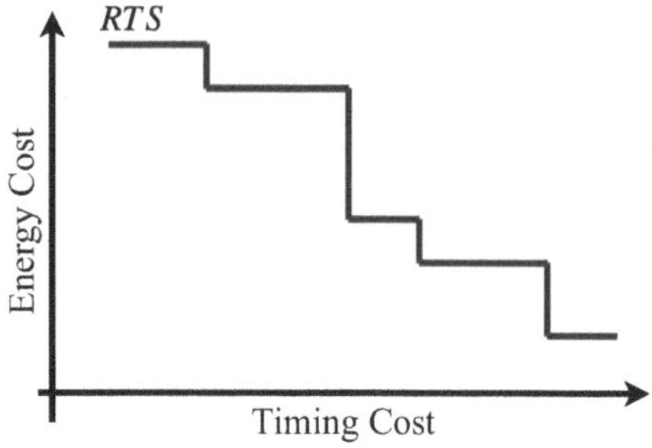

Fig. 4.1 Illustration of an RTS in the energy-time cost plain

Definition 4.1 *Thread Node* is a set of successive instructions with a deterministic behavior that is executed in a single manner, like, for example, the body of a loop. In our study, it represents the basic computational block.

Therefore, our application workload can be split into a series of TNs with their sequence depending on the control plane of the platform.

Definition 4.2 *Run-Time Situation* (RTS) is a tag/label of a TN or of a sequence of TNs, with associated primary costs, like the execution time and the energy budget.

Evidently, workload execution is a sequence of numerous RTSs. In general, a TN can be executed under different power and timing budgets, generating a number of RTSs: A processor with several DVFS points, for instance, can execute a TN in any of these steps. Hence, in the timing-energy cost plain, the RTS of a thread node can be visualized as in Fig. 4.1. Obviously, TN execution in high DVFS levels results in increased energy costs and shorter execution times.

Definition 4.3 *Cycle Budget* (N) represents the actual number of cycles of a TN (or an RTS), in the absence of any error-recovery mechanism. In the same spirit, a *Timing Budget* (T_{ref}) can be defined, that is, the reference execution time of the TN for nominal/reference frequency f_{ref}.

Moreover, we can express any kind of performance variation, and particularly variation caused by RAS interventions, in the form of *cycle noise*.

Definition 4.4 *Cycle Noise* (x) represents the extra clock cycles of a TN due to performance variation. The extra cycles lead to higher execution times ($T_{\text{real}} > T_{\text{ref}}$) of the TN when the system is operating under f_{ref}.

We have highlighted how the nondeterminism in the occurrence of hardware errors and error recovery interventions threatens timing deadlines and creates dynamic system operation. Cycle noise is injected based on system error rate (Λ), while another important parameter of x is its amplitude (M). This metric largely depends on the type of performance variation: In our case, we will focus on variation caused by RAS mechanisms. For example, an error-correcting code is a type of computer data storage technique that can detect and correct the most common kinds of internal data corruption using redundancy check codes that typically cost a few

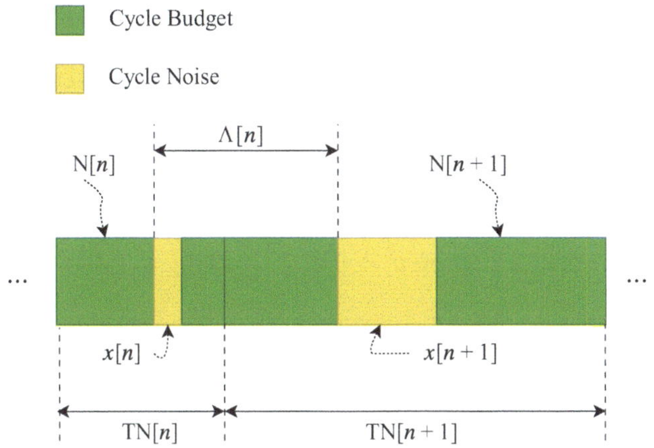

Fig. 4.2 Depicture of cycle budgets and cycle noise after TN execution

nanoseconds [1]. In contrast, a checkpoint-rollback mechanism returns the system to some correct, previous state after erroneous operations are performed and may require significantly longer timing penalties [2]. Since actual Λ and M values can be stochastic, we can also utilize λ, μ representing their respective averages over entire workload. Figure 4.2 shows a depiction of these metrics.

Definition 4.5 *Performance Vulnerability Factor* (PVF) represents the per-unit increase in the number of cycles for a specific workload due to cycle noise occurrence.

In our case, we examine the workload in the thread node level; hence, PVF after the n-th TN is formulated as follows:

$$PVF[n] = 1 - \frac{\sum_{i=1}^{n} N[i]}{\sum_{i=1}^{n}(N[i] + x[i])} \tag{4.1}$$

Definition 4.6 *Deadline Vulnerability Factor* (DVF) is the temporal equivalent to PVF and represents the per-unit increase in execution time (T) for a specific workload because of performance variation: DVF $= 1 - T_{\text{ref}}/T_{\text{real}}$. After the execution of the n-th TN, DVF is estimated according to Eq. 4.2.

$$\text{DVF}[n] = 1 - \frac{\sum_{i=0}^{n} \frac{N[i]}{f_{\text{ref}}}}{\sum_{i=0}^{n} \frac{N[i] + x[i]}{f[i]}} = 1 - \frac{\sum_{i=0}^{n} \frac{N[i]}{f_{\text{ref}}}}{\frac{\sum_{i=0}^{n-1} \frac{N[i]}{f_{\text{ref}}}}{1 - \text{DVF}[n-1]} + \frac{N[n] + x[n]}{f[n]}} \tag{4.2}$$

$f[i]$ is the operating frequency at step i. By definition, DVF and PVF are identical for constant frequency operation (i.e., $f[i] = f_{\text{ref}}, \forall i$). In practice, DVF tells us how far behind (or ahead) schedule the actual execution is, due to performance variation. An alternative metric to DVF is the slack (see Definition 4.7), which we now define in a more precise manner.

Definition 4.7 *Slack* (s) is the cumulative divergence from the cycle budget, assuming operation under reference (f_{ref}) and current ($f\,[n]$) frequency. We formulate slack according to Eq. 4.3, where $n = 0, 1, 2, \ldots, s\,[0] = x\,[0] = 0$ and $f\,[0] = f_{\text{ref}}$.

$$s\,[n] = T_{\text{ref}}\,[n] - T_{\text{real}}\,[n] + s\,[n-1] = \frac{N\,[n]}{f_{\text{ref}}} - \frac{N\,[n] + x\,[n]}{f\,[n]} + s\,[n-1] \tag{4.3}$$

The interplay between slack and DVF metrics has been proven in our previous work, verifying the equivalence of $s[n] = 0 \iff DVF[n] = 0$ [6].

4.1.2 Scenario Clustering Methodology

Modern applications typically consist of a surplus of RTSs; hence, storing their information would demand considerable resources. Moreover, a real-time identification of the current TN and its characterization to an RTS can be quite complex and with a significant timing overhead. Therefore, we follow the system scenario approach [3] and manually cluster all application RTSs into scenarios to reduce RTS-based computational complexity. This procedure takes place offline, during the profiling of the workload. Specifically, each scenario represents a bin on the time-energy plain, containing a number of RTSs with a similar behavior. Assuming that a workload is composed of k RTSs, the goal of the methodology is to limit this number into l scenarios where $l \ll k$, and now the l scenarios can easily be processed in real time. An illustration of this clustering procedure in the energy vs. timing Pareto space is presented in Fig. 4.3.

The scenario classification however comes with a sacrifice in accuracy. Specifically, the costs of the RTSs in a cluster will not be identical with those of the scenario; on the contrary, the difference in the costs is referred to as *clustering overhead* and can be visualized as the distance of the RTS from the scenario identifier in the time-energy plain. It is important to note that as the number of the scenarios increases, the overhead is minimized and higher accuracy is obtained. However, we now pay a price in complexity, since more scenarios are now generated. It is up to the developer/user to find the best trade-off and proceed with the clustering of RTSs according to his demands.

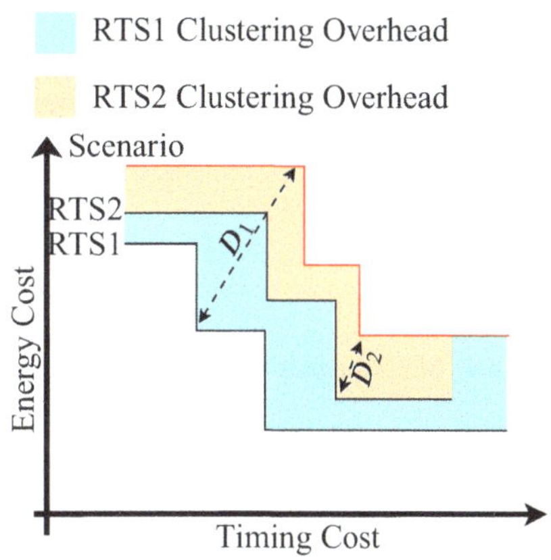

Fig. 4.3 Clustering RTSs in a Scenario. D_1 and D_2 show the distance of the RTSs from the scenario and represent the clustering overhead

4.1.3 A PID Controller Utilizing System Scenarios

In our approach, we will again utilize the PID controller with the mission to mitigate performance variations by applying suitable DVFS decisions. The controller operates at a TN-level: It first uses a *scenario detector* to identify the relative scenario cluster to which the current RTS belongs and then estimates the time slack. Therefore a design-time RTS analysis of the application under study, as well as a scenario classification, precedes the deployment of our scheme. Specifically, after the scenario clustering we develop this detector in order to categorize at run-time the current RTS with its respective scenario.

The key point in the definition of the scenario clustering is to efficiently bound the operation cost (in this case energy and time) of each RTS. Considering that each RTS has a fixed cycle budget, we have explained how several DVFS schemes provide different cost trade-offs. In our PID implementation specifically, each scenario is represented by its RTS **with the minimum delay cost**; this is important considering that scenarios are used as reference to estimate the slack in the PID controller. Through this approach, the clustering overhead creates an overestimated impression of negative slack for the other RTSs, allowing an aggressive DVFS actuation of the controller to manage performance variation. This new version of our PID controller is illustrated in the diagram of Fig. 4.4. The scenario detector receives the cycle budget of the executed RTS and decides for the relative scenario which the RTS belongs to: $\hat{N}[n]$ represents the RTS parameter of the scenario, while $\mathbf{p}[n]$ shows the actual cycle budget of the running RTS.

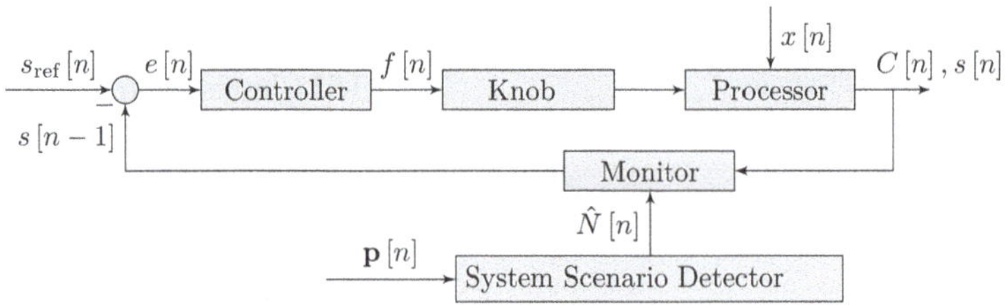

Fig. 4.4 Block diagram of the PID scheme with system scenarios [9]

4.2 Simulation Results: The PID Controller with System Scenarios

4.2.1 System Scenario Analysis

To manage dependable performance, all thread nodes have to respect their timing deadlines which are defined through their cycle budgets and the reference frequency. Since the increased TN number of the workload and the several DVFS steps of the processor generate a variety of RTSs, it is rather suboptimal to store RTS information of the workload and detect them at run-time; the number of RTSs encountered in the instruction stream of a processor is in fact significantly difficult to handle. Therefore we have highlighted the use of system scenarios to cluster at design-time the population of RTSs into cumulative representative cases. The clustering methodology for our PID controller was explained in Sect. 4.1.3.

The RTS analysis, along with a scenario clustering attempt of the spectrum sensing application, can be seen in Fig. 4.5 where each RTS represents a certain cycle budget. The DVFS points allowing us to perform the RTS analysis are shown in Table 4.1 and are based on the Intel Core2 Duo E6850 processor; we have chosen this processor for our simulation purposes since, in contrast to our target platform, detailed power and timing data are documented in prior art [7]. The multitude of solid lines in Fig. 4.5 corresponds to the application's complete analysis and already represents the complexity of handling TNs using the RTS abstraction. This

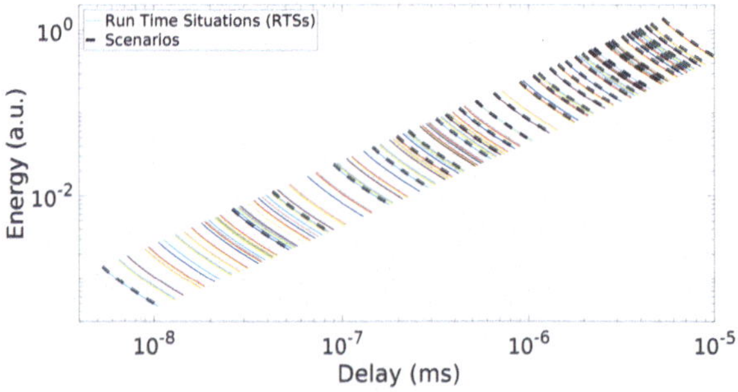

Fig. 4.5 RTS clustering to scenarios on the spectrum sensing application. Energy is calculated using the square law for dynamic power, i.e., $P_{\mathrm{dyn}} \propto f V_{\mathrm{dd}}^2$, and numbers are normalized to the smallest RTS at $V_{\mathrm{dd}} = 1.2$ V and $f = 2.588$ GHz

Table 4.1 DVFS points of the Intel Core2 Duo processor [7]

m	V_{dd} (V)	f (GHz)	m	V_{dd} (V)	f (GHz)
1.2	1.3	3.074	0.9	1.15	2.281
1.1	1.25	2.852	0.7	1.1	1.932
1.0	1.2	2.588	0.6	1.05	1.54

amount of information, evidently, needs to be abstracted further. Hence, we utilize system scenarios, which can replace the many RTSs with a more manageable set of scenarios to be used as cycle budget estimators. A close look at this figure shows how we have split the application into a pool of 768 RTSs (performing a fine-grained RTS analysis of the workload) and later grouped them into 32 scenario bins. In reality, the granularity of the RTS analysis depends on the user's/developer's demands: The final goal is to achieve the optimal RTS and scenario clustering of the workload under study.

RTSs that require a similar number of cycles to execute are classified into the same scenario. The scenario is uniquely identified with its cycle budget estimator $\hat{N}_k$, which is the minimum (or maximum in some cases) cycle budget of the RTSs that have been grouped therein. In a similar manner, RTSs with similar costs are also falling under the same scenario when perceived at run-time.[1] For our controller's scheme, the fastest RTS is assigned as the cycle budget estimator in the scenario. While moving away from the origin of the axis, new scenarios are defined, and certain RTSs are clustered to them. Eventually, the wealth of RTSs is produced by all possible combinations of the application's configuration, like, for example, accuracy settings and input size. Hence, the system scenario methodology allows us to reduce this huge set of RTSs to a more manageable set of scenarios. It is clear, however, that the complexity reduction detailed above introduces an accuracy overhead, which we have already defined as clustering overhead. There lies a trade-off between the clustering accuracy and the number of the derived scenarios: A fine-grain clustering can decrease the gap between the budget estimators; however, it increases the number of the scenarios and vice versa.

The assignment of the minimum cycle budget to a scenario forces the TNs to be more restricted, compared to their actual timing deadlines. This trend leads the PID controller to act in a more aggressive manner and boost the clock's frequency. This effect is essential for the purpose of our design: The controller's target is to ultimately absorb the interfering cycle noise of the RAS interventions. Therefore, a small additional frequency boosting operates preemptively toward the absorption of cycle noise. This "negative" clustering overhead (produced by the minimum cycle budget estimators) acts as an offset and enables the mitigation of cycle noise

[1] Note that we notate the cycle budget estimator for the n-th executed TN as $\hat{N}[n]$. Assuming that this TN is binned to scenario k, then the following equation holds $\hat{N}[n] = \hat{N}_k$. Obviously, when we use system scenarios or any other similar cycle budget estimation techniques, the estimator $\hat{N}[n]$ is used for slack calculation (Eq. 4.3), instead of the actual one $N[n]$ which in this case cannot be known at run-time.

without a reactive response. As a result, this inevitable clustering overhead is not undesirable, provided that it does not force our scheme to operate for long time periods in the “worst case” mode, which is to force it to actuate the maximum operating frequency.

4.2.2 Simulation Results

After the system scenario analysis of the spectrum sensing application, we submit this workload to the simple TN-level simulator that has been presented in the previous work [10]. The simulator operates at TN granularity: For each TN, the scenario detector provides the cycle budget estimator $\hat{N}[n]$. Moreover, we inject cycle noise based on the average values for the noise’s intensity and rate μ and λ and their respective standard deviations $\sigma\{M\}$ and $\sigma\{\Lambda\}$. The timing and power models of the simulator create traces from which DVF and total energy consumption values can be deduced. It should be noted that power models, in this case, account for both dynamic and static power consumption [7]. It is also noteworthy that the DVFS energy and timing overhead is also included in these models. After the completion of an RTS, the PID controller selects the frequency of the next iteration from the values depicted in Table 4.1. Figure 4.6 shows an instantiation of our simulator, examining the controller scheme in terms of DVF, PVF, slack, and frequency decisions.

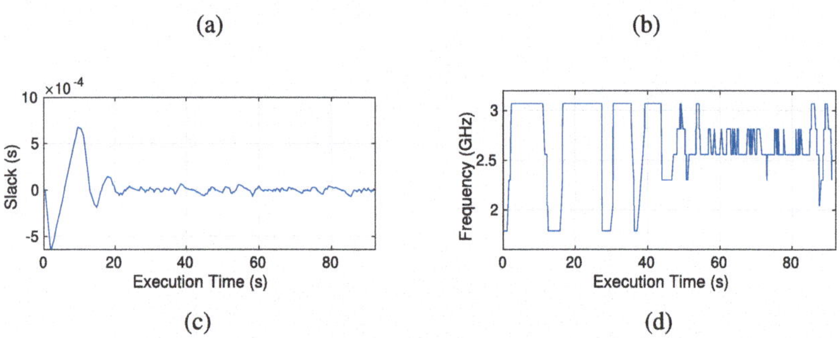

Fig. 4.6 An instantiation of our simulator examining the PID’s response utilizing system scenarios: (**a**) PVF values, (**b**) DVF response, (**c**) slack response, and (**d**) frequency decisions of the PID

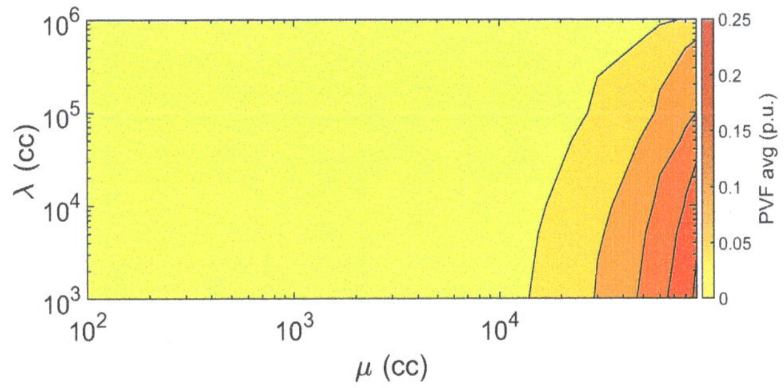

Fig. 4.7 PVF values for the set of μ and λ values

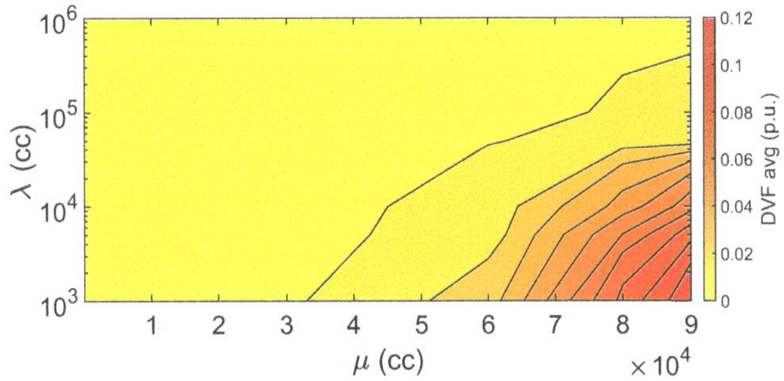

Fig. 4.8 Average DVF estimations after ten simulation iterations for the set of μ and λ values

RTSs and their cycle budget estimators are drawn from Fig. 4.5. This time we will inject cycle noise by sweeping μ and λ values and assuming $\sigma\{M\} = 0$ and $\sigma\{\Lambda\} = 0.1 \times \lambda$. This exploration will provide us with interesting results concerning the response of our control scheme across different cycle noise profiles. The simulated workload assumes equally probable RTSs. Each measurement corresponds to a stream of 10^5 TNs and, when enabled, PID controller gains are $k_p = 1.52 \times 10^{-7}$, $k_i = 5 \times 10^{-8}$, and $k_d = 4 \times 10^{-8}$. Figure 4.7 shows the PVF values for this set of noise setup.

In Figs. 4.8 and 4.9 we see average values for the DVF and energy consumption after ten simulation iterations. We can conclude that the controller manages to enable dependable performance since it absorbs the increment in PVF that cycle noise injection is causing; the challenge exists for the extreme cases of cycle noise profile (high μ, low λ). In addition, total energy consumption is also increasing as injected cycle noise moves to these profiles. This is however expected, given that, in these cases, the controller needs to work overtime. Finally, the standard deviation values of Fig. 4.10 provide insight into the statistical variability of our simulation results and indicate that the average DVF reported is substantially accurate.

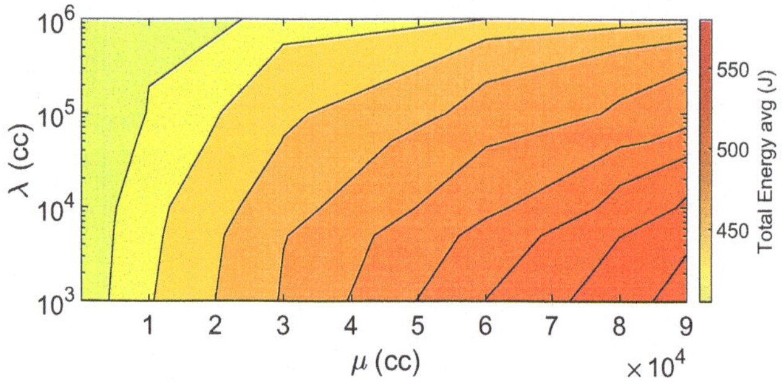

Fig. 4.9 Average energy consumption in Joules after ten simulation iterations for the set of μ and λ values

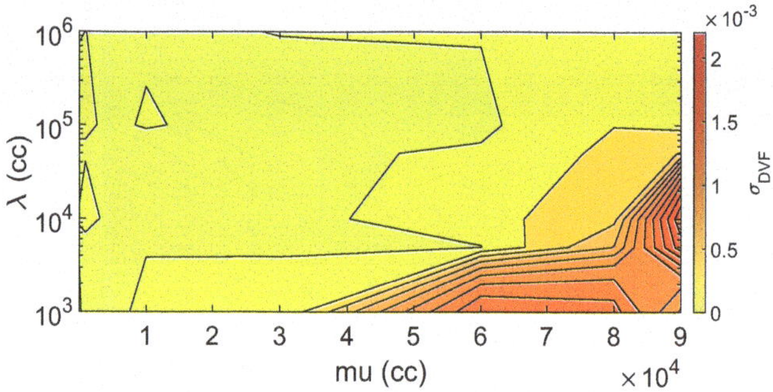

Fig. 4.10 The σ_{DVF} after the ten simulation iterations for the set of μ and λ values

4.3 Toward an Adaptive Scenario Approach

4.3.1 Scenario Adaptation: Methodology

The system scenario methodology can be easily applied to any application, given that an RTS analysis and the scenario classification are enabled at design-time. On condition that the RTS positions into scenarios can be updated at run-time, the scenario detector can modify the RTS hierarchy by re-clustering the existing scenario distribution. This process provides a run-time adjustment of the RTS' costs estimations. If an RTS is correlated with a specific cycle noise, this RTS can change its scenario reference and move to a new scenario by removing cycles equal to the cycle noise. This enables the PID controller to "adapt" to run-time conditions and absorb performance overheads before their presence. Figure 4.11 presents the basic concept behind this methodology: The RAS interventions shift the RTS position far from the start of the axes. The re-clustering process creates a positive slack that works as an offset on the expected negative slack and equalizes the time lost. The goal here is to keep the RTS original position stable. For this reason, RTS slack

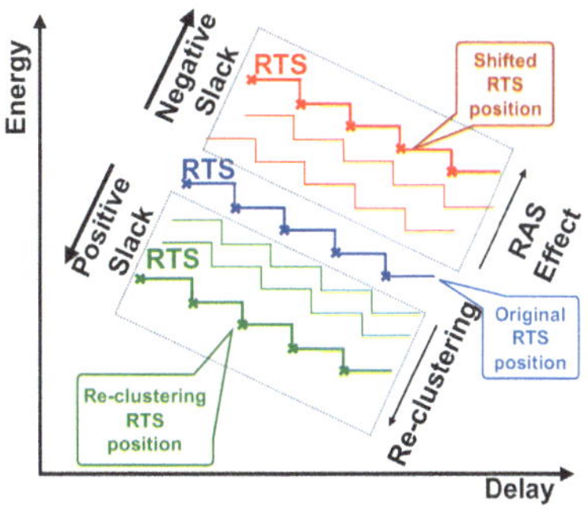

Fig. 4.11 Slack impact in the re-clustering process

history needs to be recorded and updated on fixed time windows and be exploited at run-time. The re-clustering priority focuses first on the most frequent RTSs with high slack values. This leads the PID controller to act partially proactive and consider RAS events that are gradually becoming a norm on the target processor.[2]

For each RTS repositioning, an individual re-clustering decision is taken to equalize the performance degradation. The scenario detector contains a record of the design-time scenario budgets which is updated after each re-clustering. Regarding the re-clustering decisions, the key point is to distinguish between the transient and (quasi)permanent cycle noise interference events. For example, a very brief cycle noise burst that rarely reoccurs should not be triggering a re-clustering process. The hardware degradation effects, manifesting during the operation life, cannot be predicted in advance. Thus, an updated trace history of the slack per RTS is needed to consider the error fluctuation in respect with the time. The running slack $RTS\_Slack[i]$ for a moving average window of K_w RTS executions is given by Eq. 4.4. After identifying a new trend in cycle noise events, a reformulation of the system scenarios through RTS re-clustering follows.

$$Slack_{\mathrm{RTS}[i]} = \frac{\sum_{i=0}^{N} (RTS\_Slack[N - k_w + i])}{k_w} \tag{4.4}$$

Algorithm 4.1 provides a brief description of the re-clustering process. For each RTS delay, we record the original scenario cycle budget (sc_{rts}) and the time noise (sl_{rts}) cycles. The traced RTS's slack is registered to a list ($rts\_list(rts)$); if the RTS already exists, its record is updated ($update\_rts\_slack(rts, sl_{\mathrm{rts}})$). A re-clustering is triggered if the status of the involved RTS shows a stable behavior for a specific time window ($check\_slack\_status(rts)$). Also, information about the distance from the neighboring scenarios ($Find\_scenario(Np)$) is exploited to

[2] The re-clustering process applies well when error behavior is statistical, allowing the controller to eventually adapt. On the condition that this behavior is highly random, the partial proactive nature of the scheme will be disabled.

Algorithm 4.1 RTS slack management

```
sc_rts ← detect_scenario(rts)                          ▷ Scenario identification
sl_rts ← pid_controller(sc_rts)                        ▷ Slack Identification
if sl_rts > 0 then
    if rts_list(rts) then
        update_rts_slack(rts, sl_rts)                  ▷ Update RTS slack history
        if check_slack_status(rts) then
            Np ← new_rts_pos(rts, sl_rts)              ▷ New RTS position
            Sc_fit ← Find_scenario(Np)                 ▷ Best fitting to scenario
            Merge(rts, Sc_fit)                         ▷ Merge RTS to the new scenario
        end if
    else
        store_rts_list(rts, sl_rts)                    ▷ New RTS slack registration
    end if
end if
```

recognize the optimal re-clustering decision, updating only the scenario data that have changed. Each re-clustering decision ($Merge(rts, Sc\_fit)$) is taken based on the new position of the RTS ($new\_rts\_pos(rts, sl\_rts)$) in the metric space, after reviewing the slack updated data and ensuring that the RTS deadlines in the new scenario will be respected with the minimum cost (Pareto optimal).

The re-clustering is a parallel process that does not interrupt the operation of the PID controller. In any case, the re-clustering time can easily be limited and kept as a fraction of the RTS operation time. This is due to the simplicity of the algorithm to find the new shifted positions without reordering the whole RTSs by scratch. At the end of each re-clustering, the PID controller has updated information about the suitable DVFS schemes, minimizing the generated slack while paying a price in energy. A trade-off between the produced slack and the energy exists that will be explored later along with simulation results.

4.3.2 Re-clustering Decisions

The re-clustering clearly alters the initial, design-time RTS classification by removing the default RTS reference budgets and artificially creating positive slack that shifts the position of the RTSs into the scenarios (see Fig. 4.11). This creates the impression of an extra equivalent delay to the PID controller, proportional to the removed cycles, and triggers more aggressive DVFS schemes that absolve the negative slack. Examining the re-clustering decisions that have to be considered at run-time, we outline the following cases:

- In the first case, the scenario budget is restricted enough and no slack exists; hence there is no need for scenario change (see Fig. 4.12a). The added noise (red shifted Pareto curve) triggers a revision of the RTS budget (green Pareto curve) that does not exceed the distance between the RTS and the scenario (SC_2),

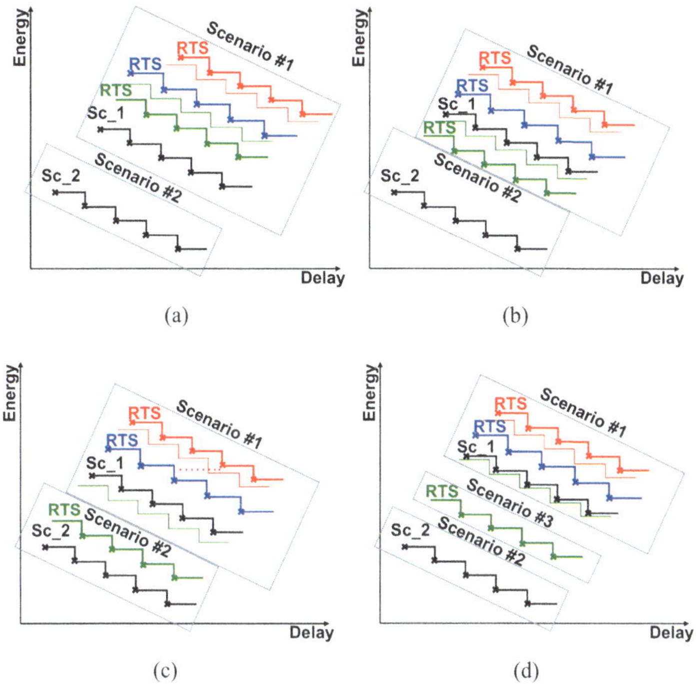

Fig. 4.12 RTS shifted cases due to RAS interventions: (**a**) RTS_1 remains in the scenario, (**b**) RTS_1 creates an individual scenario, (**c**) RTS_1 changes the scenario bounds, and (**d**) RTS_1 is merged into another scenario

exploiting the clustering overhead as an offset. Thus, clustering overhead can be utilized as a proxy of cycle noise tolerance. The next four cases cover situations where the negative slack cannot be avoided.

- In the second case (see Fig. 4.12b), in order to equalize the noise delay (red Pareto curve), the revised RTS (green Pareto curve) is located just under the scenario budget SC_1. Thus, Scenario #1 has a new more restricted case due to the RTS shifting that modifies the existing scenario budget.
- In the third case (see Fig. 4.12c), the shifted RTS (green Pareto curve) is closer to a neighboring scenario (SC_2) compared with the original (SC_1). Thus, the RTS is assigned to the new scenario (SC_2) fitting better from distance perceptive.
- In the fourth case (see Fig. 4.12d), a new distinct scenario budget is created. Two reasons lead to such a decision: (1) None of the existing scenario cycle budgets cover the new RTS constraint or (2) the creation of an individual scenario is more cost-effective by merging with an existing scenario.
- The fifth case is a combination of the second and the third aforementioned case. A neighboring scenario merges the revised RTS (as in Fig. 4.12c). However, the position of the RTS in the new scenario is located just under the scenario budget

(as the blue RTS in Fig. 4.12b) changing the new scenario cycle budget as the most restricted case.

4.4 Evaluating Adaptive Scenarios: Simulation Results

Now, we will explore the PID scheme enhanced with the adaptive scenario methodology while examining the efficiency of the new concept; for simulation purposes, we utilize once more the spectrum sensing application (see Fig. 4.5). Again, we inject cycle noise based on the values of Λ and M that represent the rate and amplitude of RAS events, respectively, in terms of clock cycles. It is assumed that cycle noise follows a bivariate Gaussian distribution with mean Λ and M values (λ and μ) and σ ($\sigma\{\Lambda\}, \sigma\{M\}$). Again, the timing and power models of the simulator create traces from which DVF and total energy consumption values can be deduced. Also, in our simulations, all overheads of DVFS actuations are included in both the timing and power models. The simulation results presented herein correspond to a set of random streams of the target application RTS, while the simulated workload assumes equally probable RTSs. To inject cycle noise, we once more sweep μ and λ values, while $\sigma\{M\}\sigma\{\Lambda\}$ are kept fixed.

The noise histogram for one stream of RTSs is shown in Fig 4.13. In fact, each measurement corresponds to a stream of RTSs, and for each combination of μ and λ values, we show the average over ten random measurements. This way, we can provide insight into the statistical variability of our simulation results. To underline the benefits of the scenario adaptivity, we compare this approach versus (a) the DVFS scheme with system scenarios, presented in Sect. 4.1.3, and (b) a (fixed) frequency guardband. A static, flat frequency usage is in fact widely employed to ensure timing deadlines [8]. To guarantee, for example, that timing constraints are not threatened by imposed error-recovery interventions or other sources of dynamism, the user can select to increase system speed in a manner that any overhead is absorbed and application finishes before its deadline. Such frequency

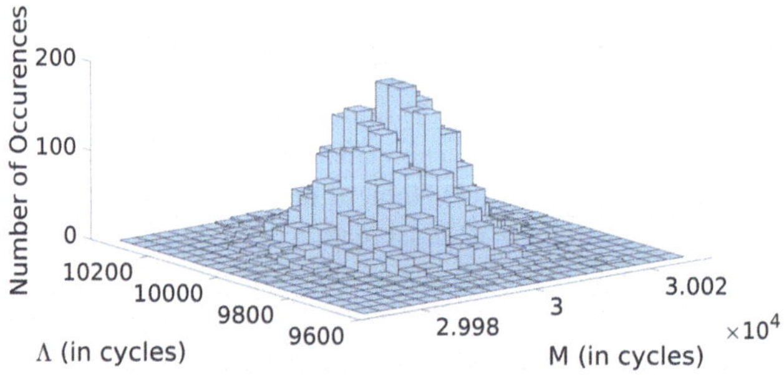

Fig. 4.13 Histogram of the cycle noise injected in one RTS

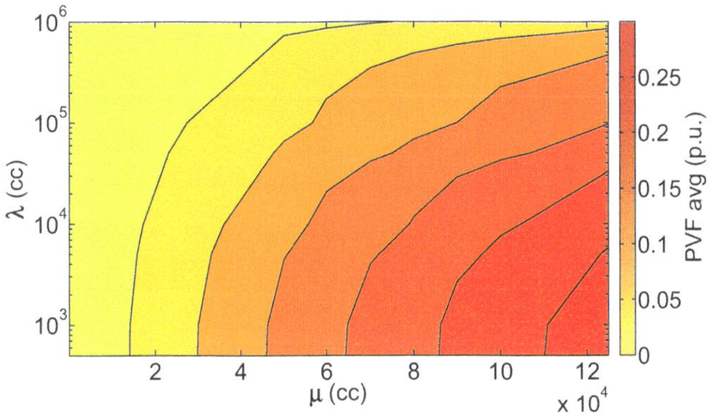

Fig. 4.14 PVF average values of ten identical iterations

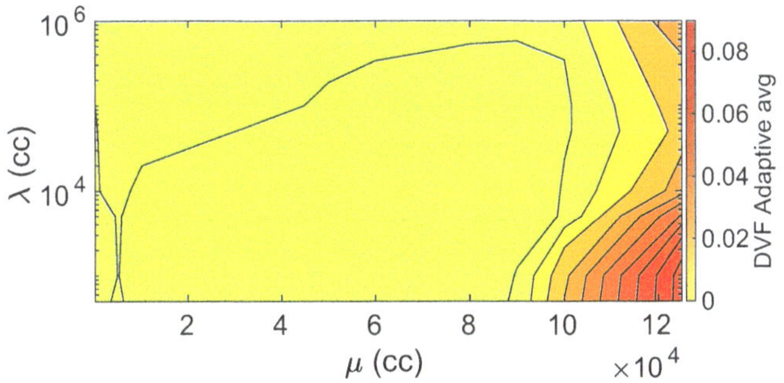

Fig. 4.15 Average DVF for adaptive scenarios after ten iterations

guardband policies, for example, can be the "performance" frequency governor of the Linux OS, where frequency is set at its highest level. From the simulation results, we can first of all conclude that, again, apart from extreme noise profiles (high μ and low λ values), the controller with the adaptive scenarios methodology manages to enable dependable performance, especially considering the increment in PVF (as seen in Fig. 4.14) which is the result of cycle noise injection.

4.4.1 Dependability Results

From Figs. 4.15, 4.16, and 4.17, we clearly distinguish that the adaptive scenarios are evidently more effective in terms of eliminating negative slack compared to the system scenarios and competitive to the guardband. In fact, the timing performance gain of the adaptive scenario approach seems to be approximately 5% higher compared to the system scenarios, while it is less effective than the gaurdband only for limited, extreme noise cases. More precisely, in Fig. 4.18 we highlight how the

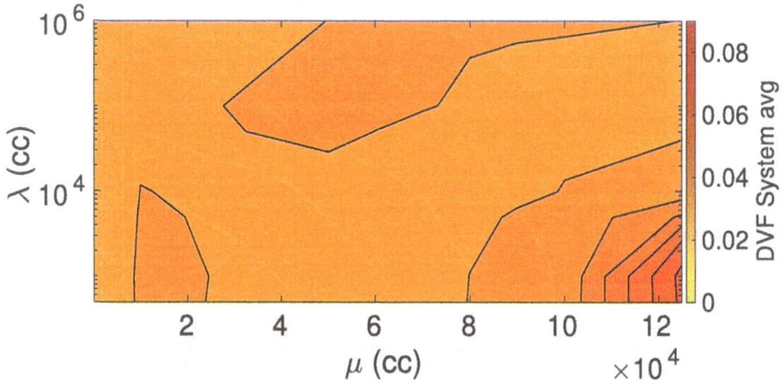

Fig. 4.16 Average DVF for system scenarios after ten iterations

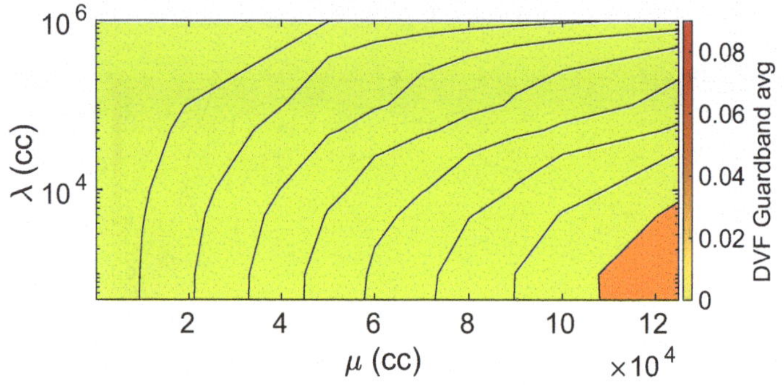

Fig. 4.17 Average DVF for frequency guardband after ten iterations

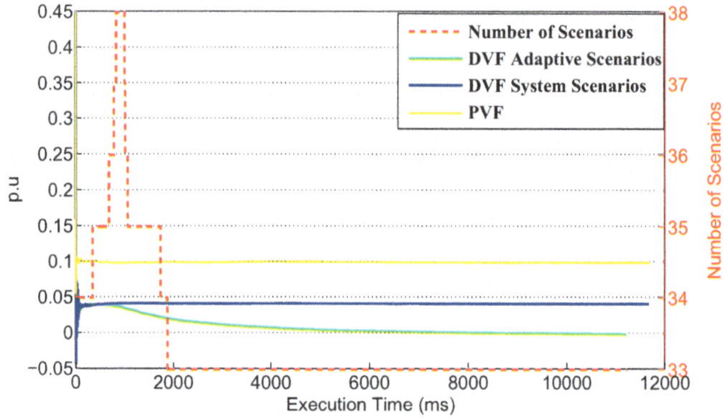

Fig. 4.18 Performance adaptability instance

adaptive scenarios clearly improve the performance dependability at a representative simulation instantiation. While system scenario DVF seems to balance to a value near zero, adaptive scenarios converge DVF closer to zero. For a certain transition period (0–2000 ms), the number of the adaptive scenarios is fluctuating to reach to a stable state. Throughout this time window, the re-clustering process creates new scenarios and gradually increases their number up to 38 (1000 ms). After this point, some of the existing scenarios are merged and reach a final number of 33 scenarios (2000 ms). Thus, the adaptive process creates and merges scenarios by applying the aforementioned re-clustering decisions and finds the optimal scenario clustering. After this process is completed (at 2000 ms), the scenarios remain stable and the DVF converges to zero without applying an over-boosting frequency policy (negative DVF) with unnecessary energy consumption results. Thus, adaptive scenarios seem to ensure performance dependability without paying a remarkable energy penalty.

4.4.2 *Energy Remarks*

From Figs. 4.19 and 4.20 we can notice that the energy gain of the two scenario methodologies is significant, compared to the guardband technology, and reaches up to 15%. This number is decreasing as the injected cycle noise profile becomes more intense (high μ and low λ). This is however to be expected, since, in this case, the scenarios attempt to mitigate the worst-case conditions that correspond to a state similar to the guardband. The energy consumption deviation between the system and adaptive scenario is negligible and does not exceed the 3%, as seen in Fig. 4.21. The reason for this is that the extra energy, consumed from frequency boosting on the adaptive scenarios, is partially equalized by the less number of DVFS switches due to the scenario adjustment that has previously taken place. Hence, we can realize that a fundamental trade-off exists between the achieved performance as expressed by the DVF and the energy impact that will be examined next.

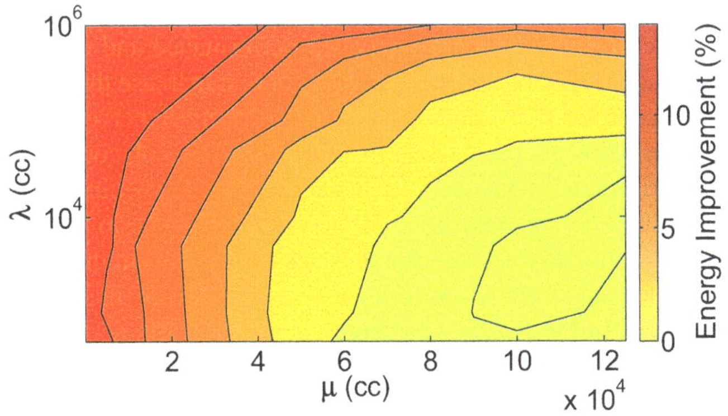

Fig. 4.19 Energy gain (%): adaptive scenarios vs. frequency guardband

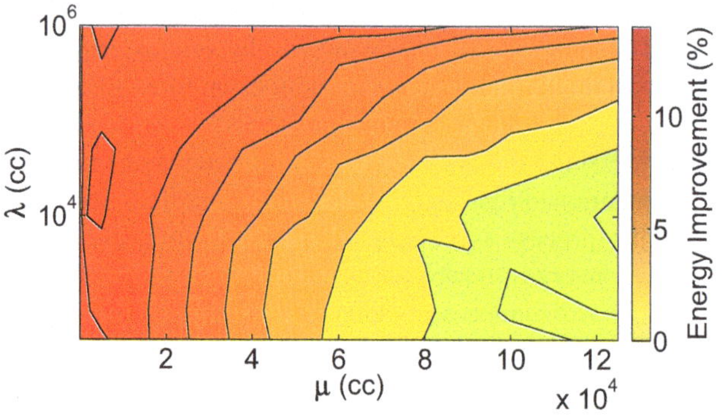

Fig. 4.20 Energy gain (%): system scenarios vs. frequency guardband

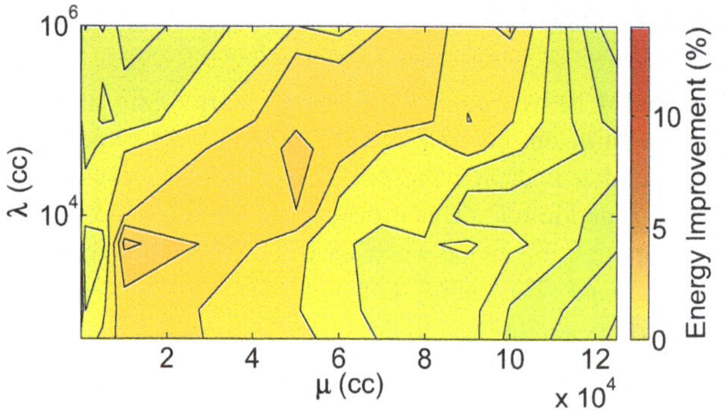

Fig. 4.21 Energy gain (%): system vs. adaptive scenarios

4.4.3 Design Trade-Offs

To understand the different trade-offs between performance and energy regarding system and adaptive scenarios approaches, we will first increase the reference cycle budget. This is completed by homogeneously applying a percentage relaxation factor F^{rel} to the MRC of each scenario (Fig. 4.3). This, of course, leads to an equal clustering overhead increase in each scenario. The new clustering overhead is formulated in Eq. 4.5. This relaxation of the reference scenario budgets provides better energy results due to the less accurate slack estimations in the PID controller. However it also presents worse performance results due to the increase of the clustering overhead. In Fig. 4.22 we present relative trade-offs between system and adaptive scenarios for several relaxation factors (0–10%) and one noise instantiation ($\mu = 5 \times 10^4$ and $\lambda = 5 \times 10^3$). We see that the Pareto curves of the two approaches seem to be very close and presenting similar cost trade-offs (delay-energy) for a specific constraint area ($0.04 < DVF < 0.08$). In fact, the adaptive scenarios approach appears just a little worse trend (longer distance from the start of the metric

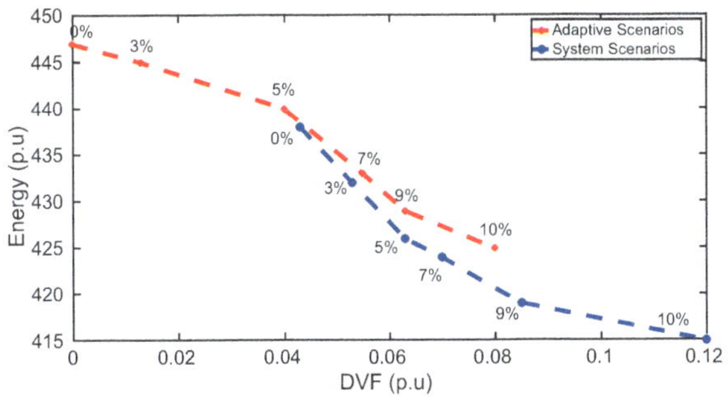

Fig. 4.22 Performance (DVF) vs. energy trade-offs

axes) because of the extra, required implementation cost. We can realize that the adaptive scenario approach is of high value when strict dependability constraints are essential; clearly, the static approach is unable to meet such demands. Nevertheless, when timing constraints relax ($DVF > 0.04$), the static scenarios scheme appears to be Pareto-optimal and should be utilized instead.

$$Cl'_{\text{RTS}[i]} = \sum_{i=0}^{n} (1 + F_{\text{rel}}) \times MRC_{\text{RTS}[i]} - OP_{\text{RTS}[i]} \tag{4.5}$$

4.5 Conclusion

In this chapter, we have briefly presented the theoretical background on system scenarios while proposing, in Sect. 4.1, an enhanced version of the PID controller, which actuates DVFS on thread node granularity. Next, in Sect. 4.2, we have verified the operation of our scheme using several cycle noise profiles. For simulation purposes, we use the Intel Core2 Duo E6850 processor which had sufficient and detailed data, documented in prior art. Section 4.3 discusses our adaptive scenario approach that allows the re- clustering of the scenario hierarchy at run-time. We elaborated on the re-clustering methodology while studying, in detail, relative re-clustering decisions. Finally, in Sect. 4.4, we have simulated a PID implementation working with adaptive scenarios and compared the results versus:

1. A controller working with system scenarios and
2. A frequency guardband policy

Simulation results have proven the efficiency of the adaptive scenarios scheme in terms of both dependability management and energy consumption. However, the full timing guarantees are not ensured in these schemes since the probability of deadline misses is not completely eliminated both in theory and in practice. To ensure full timing guarantees, dynamic scenarios need to be introduced, which allows the prediction and identification of the scenarios of future workload. Therefore, in the next chapter we will further propose and discuss about the mitigation approaches instantiating dynamic scenarios, which can truly ensure timing guarantees.

References

1. Amato, P., Bellini, S., Ferrari, M., Laurent, C., Sforzin, M., Tomasoni, A.: Fast decoding ECC for future memories. IEEE J. Sel. Areas Commun. **34**(9), 2486–2497 (2016)
2. Chan, C., Schwartz-Narbonne, D., Sethi, D., Malik, S.: Specification and synthesis of hardware checkpointing and rollback mechanisms. In: DAC Design Automation Conference 2012, pp. 1222–1228 (2012)
3. Gheorghita, S.V., Basten, T., Corporaal, H.: Application scenarios in streaming-oriented embedded-system design. IEEE Design Test Comput. **25**(6), 581–589 (2008)
4. Gheorghita, S.V., Palkovic, M., Hamers, J., Vandecappelle, A., Mamagkakis, S., Basten, T., Eeckhout, L., Corporaal, H., Catthoor, F., Vandeputte, F., Bosschere, K.D.: System-scenario-based design of dynamic embedded systems. ACM Trans. Des. Autom. Electron. Syst. **14**(1), 3:1–3:45 (2009)
5. Hardy, D., Puaut, I., Sazeides, Y.: Probabilistic WCET estimation in presence of hardware for mitigating the impact of permanent faults. In: 2016 Design, Automation Test in Europe Conference Exhibition (DATE), pp. 91–96 (2016)
6. Noltsis, M., Rodopoulos, D., Zompakis, N., Catthoor, F., Soudris, D.: Runtime slack creation for processor performance variability using system scenarios. ACM Trans. Des. Autom. Electron. Syst. **23**(2), 24:1–24:23 (2017)
7. Park, S., Park, J., Shin, D., Wang, Y., Xie, Q., Pedram, M., Chang, N.: Accurate modeling of the delay and energy overhead of dynamic voltage and frequency scaling in modern microprocessors. IEEE Trans. Comput.-Aided Design Integrated Circuits Syst. **32**(5), 695–708 (2013)
8. Pering, T., Burd, T., Brodersen, R.: The simulation and evaluation of dynamic voltage scaling algorithms. In: Proceedings. 1998 International Symposium on Low Power Electronics and Design (IEEE Cat. No.98TH8379), pp. 76–81 (1998). https://doi.org/10.1145/280756.280790
9. Rodopoulos, D.: Modeling and Mitigation of Parametric Time-Dependent Variability in Digital Systems. KU Leuven. Faculteit ingenieurswetenschappen, Leuven (2016). https://lirias.kuleuven.be/handle/123456789/529875
10. Rodopoulos, D., Weckx, P., Noltsis, M., Catthoor, F., Soudris, D.: Atomistic pseudo-transient BTI simulation with inherent workload memory. IEEE Trans. Device Mater. Reliab. **14**(2), 704–714 (2014)

Chapter 5
Workload-Dependent Multi-Timescale Mitigation Approach for Performance Variability: Theory

Abstract In Chap. 4, effective mitigation approaches for performance variability based on concepts of system and adaptive scenarios have been presented. However, the goal of full timing guarantees has not yet been achieved in these approaches. Therefore, here we further propose a novel workload-dependent multi-timescale mitigation approach to ensure timing guarantees for highly dynamic workloads, tackling the variability from the architecture and software layers in sub-millisecond time granularity. As motivated in Chap. 3, the specific problem formulation of mitigating performance variability with full timing guarantees has not been achieved up to now in state-of-the-art. Nevertheless, our approach will show that full timing guarantees can be achieved by workload-dependent approaches. In the following two chapters, we present the approach in full detail. In this chapter, the principles and design of the experimental approach are elaborated, while in Chap. 6 the simulation we performed in testing this approach is discussed.

This chapter is structured as follows: Sect. 5.1 describes the motivation and design principles of this approach. The rest of this chapter introduces the salient features of this approach. In Sect. 5.2, we present heterogeneous datapaths (HDPs), a novel system knob. Section 5.3 shows how multiple knobs can collaborate to deal with the variability of different timescales. Thereafter, the complete scheduling algorithm based on dynamic scenarios is elaborated in Sect. 5.4. Then, we further provide the proof that the algorithm can ensure timing guarantees in Sect. 5.5. Finally, the applicability of this approach to other architectures and technologies is discussed in Sect. 5.6.

5.1 Motivation and Principles

As mentioned in Sect. 3.3.2, ordinary mitigation approaches for time-critical and time-sensitive applications are composed of a DVFS knob and an algorithm based on the WCET model. Although these approaches provide necessary timing guarantees in typical cases, they show strong drawbacks and limitations which undermine their practical application in providing full timing guarantees in highly dynamic workloads. Hence, the state-of-the-art has to fall back to some type of improved

J.-Y. Lin et al., *Workload Dependent Mitigation Approaches for Performance Variability*, https://doi.org/10.1007/978-3-031-85624-2_5

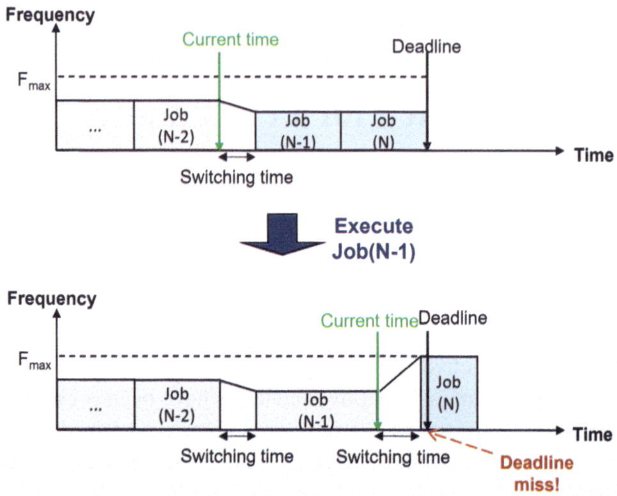

Fig. 5.1 Example of how a deadline miss could occur if switching time and job execution time are comparable. The unexpected long execution time of Job $(N-1)$ and the long switching time leave almost no time budget for Job N, resulting in the deadline miss

best-effort approach where the number of violations of the timing specification will occasionally exceed the imposed bound.

The first limitation of the mitigation approaches with timing guarantees lies in the WCET model . As discussed in Sect. 3.3.2, the WCET model is too simplistic to reflect the sophistication of modern hardware and computer architectures, resulting in over-margining and failed timing guarantees. Thus, workload-dependent approaches are the only viable option for potentially still ensuring timing guarantees. However, to the best of our knowledge, no known approaches have been published yet which achieve full timing guarantees.

Second, the switching time of DVFS limits the time granularity of mitigation to the millisecond level, as discussed in Sect. 3.2.1. In the scheduling of a sequence of jobs, the switching time of system knobs must be significantly shorter than the execution time of individual jobs. Otherwise, long switching time could make the deadlines difficult to meet. Figure 5.1 demonstrates an example of how a deadline miss could occur if switching time and job execution time are comparable. The upper and lower parts of this figure show the assumed schedule before and after the execution of Job $(N-1)$, respectively. Job $(N-1)$, a job close to the deadline, has an exceptionally long execution time, out of the expectation of the controller. And due to the long switching time, there is almost no time budget left to execute the Job N. Therefore, the deadline is missed even though the Job N is executed under the fastest mode. To avoid the deadline miss, the controller can increase the speed for the previous jobs so that Job $(N-1)$ finishes much earlier to leave enough timing budget for Job N plus the switching time. However, the energy consumption would be increased significantly.

Moreover, the switching time reduces the computation time spent on the original workload, consequently decreasing the maximum achievable throughput and increasing the speed and energy consumption required to execute the same workload. Therefore, the switching time should be limited to an amount that does not significantly influence the energy consumption and the throughput of the workload. If we assume the reasonable switching time overhead to be at most 1% of the original execution time, then in the case of DVFS, the minimum job execution time needs to be several milliseconds. The constraint in the job length prevents the mitigation from operating in finer granularity in time, degrading its capability to tackle sub-millisecond variability.

Our novel workload-dependent multi-timescale mitigation approach is developed to overcome the aforementioned drawbacks. This approach aims to provide guarantees in different conditions where applications are required to meet certain timing specifications. A typical specification is periodic deadlines in real-time applications, such as multimedia applications. Moreover, in general applications, processors are expected to process users' commands and respond to users within a maximal latency, so that the user experience is not degraded. These different requirements of timing guarantees can be handled in a unified way in our approach.

The proposed approach aims to deal with the performance variability from the architecture and software layers. As discussed in Sects. 2.2 and 2.3, different inputs of the software can lead to different execution flows in the architecture and the software, causing variability in execution time. In essence, the execution time of running certain software with given inputs on architecture is deterministic. In other words, it is possible to calculate the exact execution time if we know every detail of the architecture, software, and workload. However, this calculation is too complicated to be feasible at run-time. Therefore, workload-dependent variability models of execution time need to be created for the use of run-time execution time prediction and bounding. We suppose the best form of the variability model is the dynamic scenario-based statistic table generated according to design-time profiling analysis, which we will describe later in Sect. 5.4 and demonstrate by two example applications in Sects. 6.2.1 and 6.3.1.

Figure 5.2 shows the block diagram of the workload-dependent multi-timescale mitigation approach. In general, this approach is built on the principle of a performance variability mitigation system as described in Sect. 3.1, consisting of a run-time controller, system knobs, and timers as monitors of performance variability. The controller runs in the optimized firmware or a standalone power management unit, so that the time overhead is minimized. On top of that, this approach has two main features, which improve the two aforementioned drawbacks in state-of-the-art mitigation approaches.

First, the new HDP knob is introduced. With its short switching time of tens of nanoseconds, HDP improves the time granularity of mitigation to the microsecond level. Furthermore, we develop a run-time controller of collaborating multiple knobs, so the merits of both HDP and DVFS can be exploited for mitigation. Second, a workload-dependent algorithm that properly instantiates the dynamic scenario concept is developed. It can ensure timing guarantees even when a highly dynamic

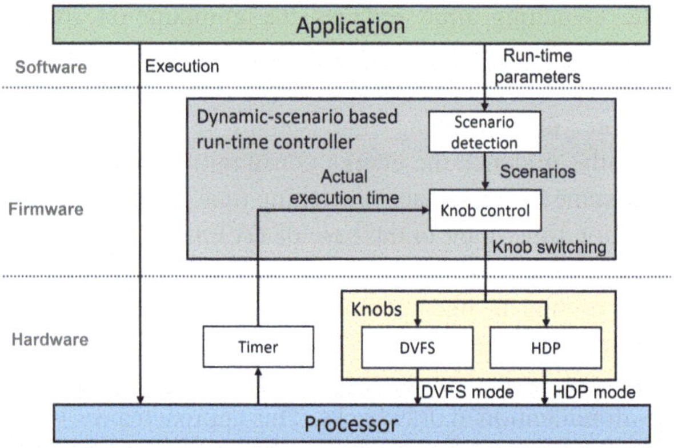

Fig. 5.2 Block diagram of the workload-dependent multi-timescale mitigation approach

workload is executed. Thanks to the two features, this approach can proactively respond to highly dynamic performance variability down to the microsecond level to ensure timing guarantees.

5.2 Heterogeneous Datapaths

Heterogeneous datapaths (HDPs) are multiple datapaths of the same functions in the same processor, each with different speed, power, and energy consumption. The heterogeneity of datapaths can originate from differences in circuit design, V_{th} of devices, body bias, etc. At run-time, the processor can dynamically select which datapaths to activate so that its requirements of speeds and energy consumption are fulfilled. In the past research, datapaths of multiple modes have been proposed, bearing similar concepts of HDP. Some examples include variable-latency functional units [10] and heterogeneous functional units in superscalar processors [5]. Nevertheless, the goals of these datapaths are to minimize average latency or maximize throughput. To our knowledge, HDP is the first to extend the use of multiple datapaths as a system knob enabling run-time changing of speed and energy levels.

To instantiate this idea, we implement HDP design in the RISC-V Ariane core [14], which is a modern representative 64-bit processor core. Ariane core is a building block of OpenPiton [1], a RISC-V multicore processor platform that is available through open access. Figure 5.3 shows the microarchitecture of the Ariane core, which consists of six pipeline stages. It is an in-order (in-order issue, out-of-order execute, and in-order commit), single-issue, 64-bit application-class processor. The dynamism in the microarchitecture includes out-of-order execution, branch prediction, and both instruction and data caches. With these features, it

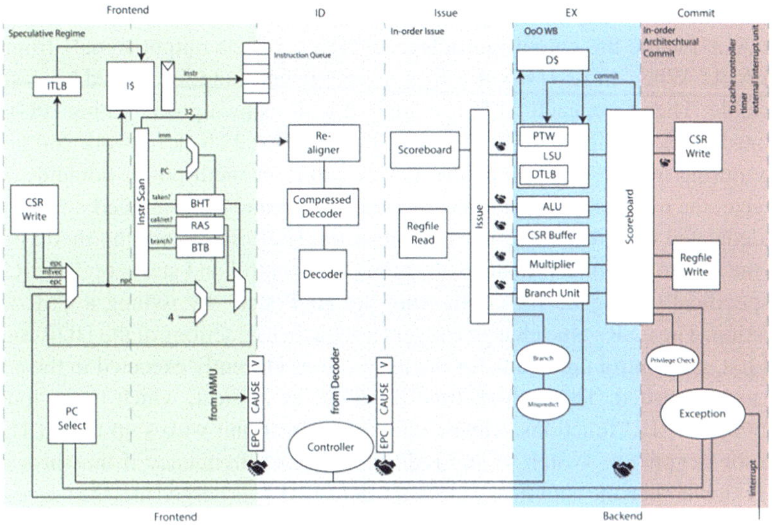

Fig. 5.3 Block diagram of the Ariane core. Replotted from [14]

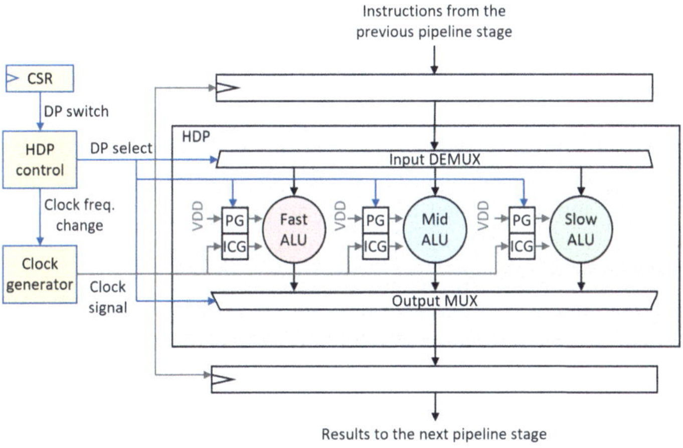

Fig. 5.4 Schematic of the HDP design in the ALUs of Ariane core

is a representative processor core which contains sources of architectural-layer performance variability.

We equip the Ariane core with HDP in multiple units of the execution stage, including the ALU, CSR buffer, multiplier, and branch unit (the collection of these units is referred to as ALU in the rest of this book for simplicity). Our HDP design is demonstrated in Fig. 5.4, in which three ALUs (fast ALU, mid ALU, and slow ALU) of the same function but different clock frequencies and energy levels are implemented in the pipeline stage of execution. At run-time, only one ALU is active at one time. The input demultiplexer (DEMUX) dispatches input signals to the active ALU, while the output multiplexer (MUX) collects output signals from the active ALU. When an ALU is not active, its energy consumption should be as small as possible. Therefore, each ALU is equipped with its own power gating (PG) and integrated clock gating (ICG) cells to shut down its local clock network and power supply during its idle time. To enable power gating in small power domains of an ALU size, the microarchitectural power gating technique [7] is applied.

A dedicated circuit is designed to control the HDP. To switch the mode of the HDP, the controller simply changes the value of a control and status register (CSR) [12] specifically assigned for configuring the HDP state, by issuing a CSR-write instruction. The CSR value change triggers a sequence of actions of the HDP control unit. First, the control unit waits for the instructions currently executed in the active ALU to be finished. The waiting time is 66 cycles at most, which is the longest latency of all ALU functions. Meanwhile, the control unit wakes up the ALU, the processor is going to switch to and reduces the clock frequency if the successive ALU is slower than the active one. Once the active ALU is purged and the successive ALU is ready, the control unit switches the modes of the input DEMUX and the output MUX to change the direction of data signals. Finally, the control unit enables clock gating and power gating of the previously active ALU and raises the clock frequency if the successive ALU is faster than the previously active one.

The switching time of the HDP , including the waiting time, the changing time of the clock frequency, and the wake-up time of power gating, only adds up to tens of cycles in total, equivalently tens of nanoseconds in a normal processor of several GHz. The switching time overhead can be kept under 1% even if the HDP switches every tens of microseconds. Therefore, HDP can serve as a knob of fine-grained mitigation at the sub-millisecond level. For all practical applications with timing guarantees that we are aware of, this limit of μs granularity is sufficient for the imposed timing specifications.

The HDP is implemented based on the 3 nm technology of IMEC [13], demonstrating the feasibility of the approach for highly scaled modern technologies where the variability ranges are the highest and providing the timing guarantees is hence the hardest. Each ALU is synthesized and physically designed with different clock frequency constraints. The relations of the clock frequency and energy of physical designs of ALUs are shown in Fig. 5.5, simulated with a JPEG decoding workload. Different V_{th} options of devices contribute to the differences in the speed and energy among designs. In principle, ALUs of higher speed and lower energy are composed of devices of lower V_{th}. Designs with multiple device V_{th} types are generally better in the speed-energy trade-off, since they can be optimized by using lower-V_{th} devices in critical paths and higher-V_{th} devices in noncritical paths. The circuit topology also contributes to the differences in the speed and energy.

The candidates of ALU designs forming the HDP should be chosen from the ones on the Pareto front of energy-speed relations. To realize the full potential of the mitigation, the HDP design should make the most of the speed-energy curve. This means that the fastest HDP mode should be as fast as possible before the point at which speed improvement is so small that additional energy investment is not

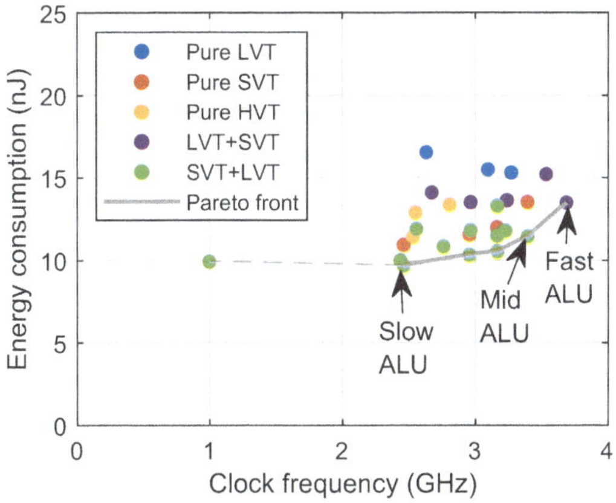

Fig. 5.5 Energy consumption vs. clock frequency of physical designs of ALU

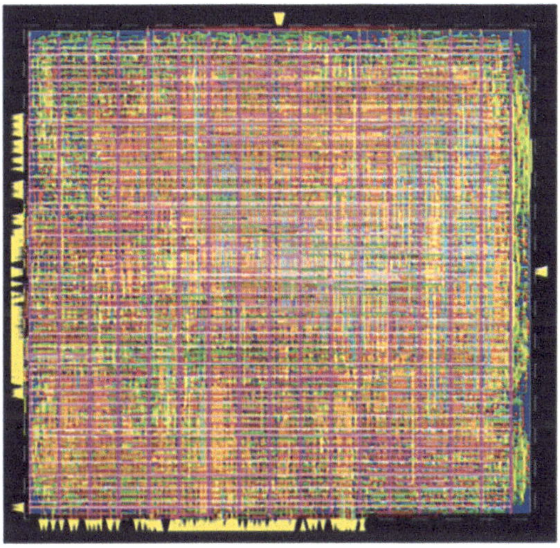

Fig. 5.6 Layout of the fast ALU of the HDP

worth it any more. On the other hand, the slowest HDP mode should be as slow as possible until the point at which the energy is flattened and even increases due to larger leakage energy brought by longer execution time.

As a result, our HDP design has three modes, each of which uses a different ALU design on the speed-energy curve in Fig. 5.5. Figure 5.6 shows the layout of the physical design of the fast ALU, and Table 5.1 summarizes the designs of ALUs in the HDP. Each ALU contains about 12,000–19,000 gates, within which about 800–1200 are flip-flops. To diversify the speed and energy of ALUs, devices of different V_{th} are used. Compared to the fast ALU, the mid ALU saves 15% energy by 8% speed degradation, while the slow ALU saves 28% energy by 33% speed degradation. The design optimization of HDP by state-of-the-art EDA tools is limited due to the complexity of the timing constraints, so we expect the energy efficiency of HDP could be better in future with the improvement of EDA tools.

Table 5.1 Summary of the ALU designs in the HDP

Design	Fast ALU	Mid ALU	Slow ALU
Clock frequency	3.69 GHz	3.39 GHz	2.46 GHz
Number of total gates	18099	18092	12827
Number of flip-flops	1188	1188	853
Number of other gates	16911	16904	11974
Multi-V_{th} libraries used	LVT, SVT	SVT, HVT	SVT, HVT
Technology	3 nm [13]		
Operating condition	TT/0.7V/25C		

The introduction of HDP also induces overhead on the chip area . The amount of the area overhead depends on the number of modes of the HDP. Since the hardware of HDP is essentially multiple functional units of the same functions in different circuit topologies, the chip area of HDP is approximately multiple times the total area of the original functional units. However, in modern processors, the functional units only account for a small proportion of the whole chip area. Therefore, the area overhead is relatively small compared to the whole chip. In our HDP design, the total three ALUs take up approximately 2.9x the area of the original fast ALU. However, in the original Ariane design, the ALU and the multiplier only account for 13.2% area of the core, while the area of the processor core is estimated to be 10% of the whole area of the SoC [14]. Therefore, the area overhead of the HDP is estimated to be only 2.5% of the SoC.

While we demonstrate the HDP design only in the example of the Ariane datapath, the same design principle are viable for other microprocessor cores and all kinds of datapaths. It can also be easily extended to domain-specific architectures, such as GPUs or highly optimized parallel matrix-multiply-accumulate accelerators in the field of machine learning. Apart from datapaths, the concept of bringing heterogeneity into microarchitecture is also potentially applicable to other components in the entire macroarchitecture. For the memories, switching among different types of memory instances which exhibit different delay-energy trade-offs is quite easy to achieve. For example, one can imagine having part of a memory macro with smaller subbanks to store time-critical data. We look forward to the extensions of this concept to other components (interconnect networks, etc.) to be explored in the future, so that the energy efficiency and resiliency of the individual components and the entire macroarchitecture can be enhanced.

5.3 Collaboration of Multiple Knobs

Variability of different sources occurs in different timescales. For example, as mentioned in Sects. 2.1.2 and 2.1.3, abrupt voltage droop can happen in tens of nanoseconds, while temperature normally changes gradually over hundreds

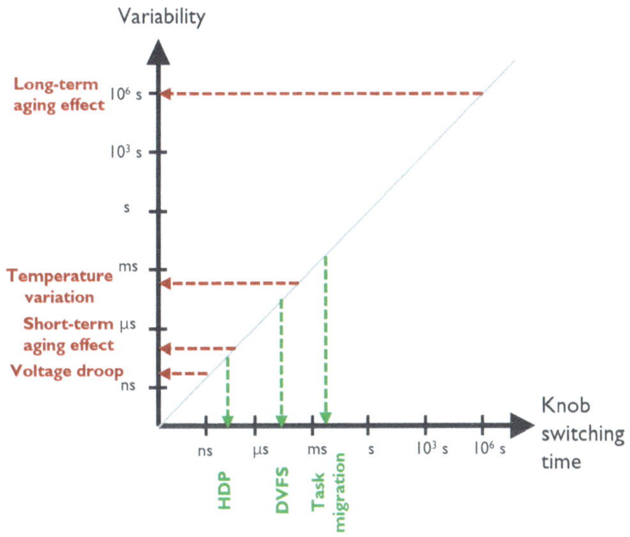

Fig. 5.7 Variability sources and mitigation knobs arranged by the timescales, which tells whether the knobs can react faster than the variability for successfully mitigation

of microseconds. Depending on characteristics, architectural and software-level variability spans over a wide range of timescales from nanoseconds to seconds. To mitigate a kind of variability, the switching time of a knob must be shorter or comparable to the timescale of the variability. Otherwise, the knobs will not be quick enough to react to the variability.

A holistic view of the timescales of variability sources and mitigation knobs is illustrated in Fig. 5.7. The timescales of the knobs make them suitable for tackling a different type of variability sources. For example, the 10–150 μs switching time of DVFS makes it effective in tackling temperature fluctuation, but too slow to react to voltage droop. Therefore, a complete solution for all kinds of performance variability should integrate multiple knobs of different switching times, so each kind of variability is handled by the knobs of its respective timescales.

In our approach, the multi-timescale concept is embodied in the use of both HDP and DVFS. These two knobs are complementary in switching time, speed range, and energy range. DVFS enables wide speed and energy ranges, but its switching time is longer, so it is suitable to react to the large and slow variability at the millisecond level. On the other hand, HDP can switch in a shorter time and provide fine-grained speed and energy levels, but its speed and energy ranges are limited, so it is suitable to respond proactively to small and abrupt variability.

To illustrate this concept, Fig. 5.8 shows all available speed and energy levels of executing a JPEG-encoding workload on the ALUs of the Ariane core equipped with both HDP and DVFS, benchmarked with the cycle-accurate simulation flow described later in Sect. 6.1.1. These levels are fine-grained and cover wide speed

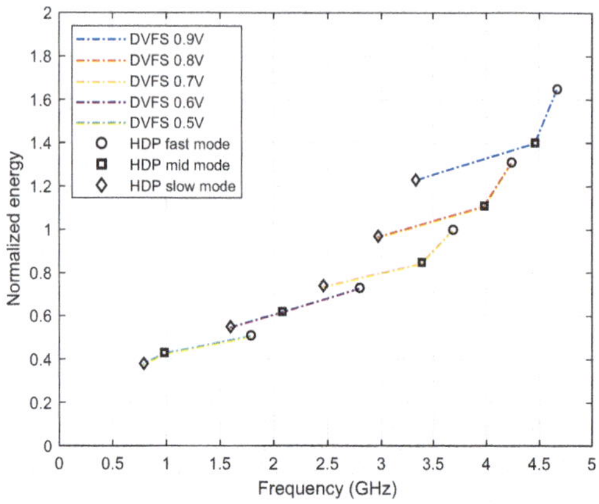

Fig. 5.8 Speed and energy trade-off of the ALUs running in each of the possible combinations of the DVFS mode (0.5 V–0.9 V) and the HDP mode (fast, mid, and slow)

and energy ranges, in which the slowest mode has only 17% speed and 23% energy consumption compared to the fastest mode.

In general, the modes of better energy efficiency (low energy-to-speed ratio) are preferred. From Fig. 5.8, one can obtain the Pareto-optimal modes in terms of energy efficiency, which are the modes close to the lower right corner of the figure. Although several modes are not Pareto-optimal, fast switching time could allow these modes to have more energy saving in mitigating fast variability. For example, switching from the 0.7/mid mode to the 0.6/fast mode requires switching time overhead of both DVFS and HDP, which reduces the time spent in the latter mode, lowering the energy efficiency. In this case, switching from the 0.7/mid mode to the 0.7/slow mode can yield better energy efficiency in the short term due to shorter switching time, even though the 0.7/slow mode is not Pareto-optimal.

The design of the knobs should consider the effectiveness and flexibility of the mitigation approaches. In principle, a good set of modes should have high energy efficiency, wide ranges of speed and energy, and fine-grained speed and energy levels. One can imagine a fully optimized set of modes to consist of Pareto-optimal modes in terms of energy efficiency, in which the neighboring modes are separated by a minimal interval of speed and energy levels. However, the optimization of knobs is restricted by the circuit design and the underlying technology. For example, the design of HDP is limited by the design space of circuit topology and the device V_{th} flavors. Across this design space, the range of speed and energy that yields good energy efficiency is relatively small, as shown in Fig. 5.5. In fact, most of the range of good energy efficiency is covered by the modes of our HDP design. Moreover, since the same HDP hardware is used regardless of the DVFS setting, optimizing a particular HDP curve in Fig. 5.5 will also affect the other HDP curves, potentially making other curves suboptimal. Therefore, it is not possible to optimize each mode individually at the same time. Due to the nonideality of the knobs in switching time, speed/energy range, granularity, and energy efficiency, it is crucial for the scheduling algorithm to smartly schedule the mode setting to minimize these drawbacks for energy minimization and timing guarantees.

5.4 Dynamic Scenario-Based Scheduling Algorithm

The adaptive control algorithm is based on the principles of dynamic scenarios (DSs), aiming to proactively respond to performance variability to ensure timing guarantees. In principle, the algorithm takes advantage of the information of the dynamic scenarios, which are the groups of workload segments with similar execution time and energy costs, defined in design-time profiling. By identifying the scenario that the system encounters at run-time, the algorithm can predict execution time with limited error and find an upper bound of the execution time which is substantially less pessimistic than the WCET model.

In a nutshell, the core of dynamic scenarios inherits the system and adaptive scenarios, sharing similar principles in realizing design-time/run-time co-optimization methodologies that tackle any kind of dynamism in modern hardware/software systems with low design margins. The DS principles are developed in previous research and elaborated in [3, 4]. These generic principles can be instantiated into practical approaches that tackle different variability problems in every new context.

Here, we instantiate the DS principles in the novel context of mitigating timing variations, which results in a new mitigation approach tuned to our specific objectives of ensuring timing guarantees for processors equipped with speed-varying knobs like HDP and DVFS. The DS-instantiated approach consists of a design-time scenario clustering method for creating dynamic scenarios, as well as a run-time scheduling algorithm for selecting the optimal mode of the knob while ensuring timing guarantees. The concepts and the instantiation of the dynamic scenario principles are explained in the following sections, while two implementations for different applications are shown in Sect. 6.2 and 6.3.

5.4.1 Principles of Dynamic Scenarios

In principle, dynamic scenarios , similar to system scenarios and adaptive scenarios, are groups of similar system behaviors from the perspective of multidimensional system costs, such as execution time, energy consumption, and resource requirements. The scenarios provide a reference by which the run-time controller can estimate or calculate a bound of the system costs of the upcoming workloads. By changing the system behaviors to adapt to the upcoming scenario at run-time, design systems can improve performance, reduce costs, and possibly recalibrate themselves during execution. The principles of dynamic scenarios are applicable in many contexts, instantiated to become solutions to many run-time system optimization problems, such as DVFS scheduling, energy-efficient exascale computing, and flexible manufacturing system design [3].

Figure 5.9 depicts the basic concepts of dynamic scenarios. Based on the design-time profiling and control flow analysis, the workload is segmented into atomic units called *run-time situations*, as depicted in Fig. 5.9a. Each RTS is a situation of

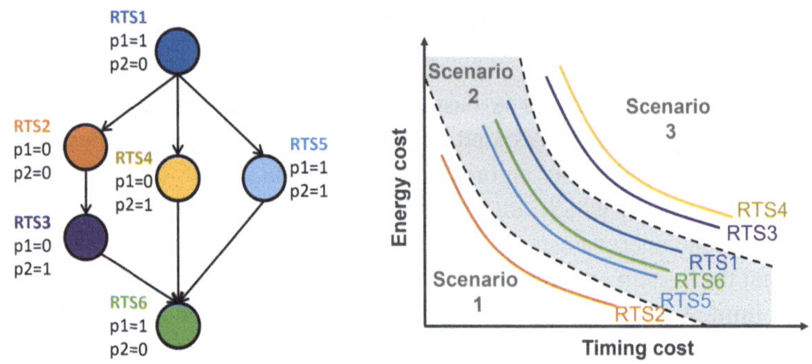

(a) (b)

Monitoring parameters		Scenario	RTS	System costs	
p1	p2			Timing	Energy
0	0	**1**	2	low	low
1	0, 1	**2**	1, 5, 6	middle	middle
0	1	**3**	3, 4	high	high

(c)

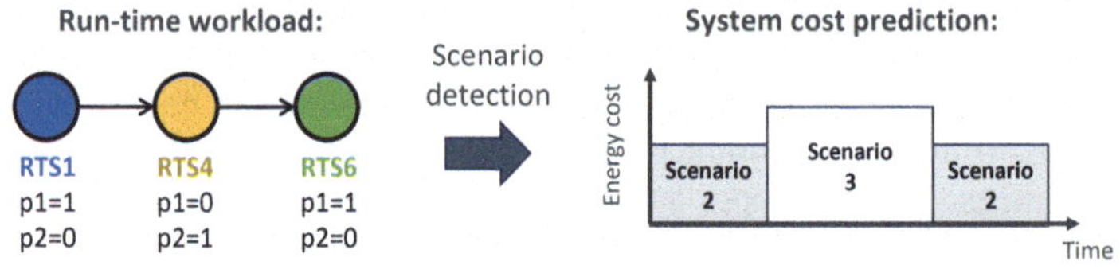

(d)

Fig. 5.9 Depictions of the basic concepts of dynamic scenarios. (**a**) At design time, different parts of the workload are classified into run-time situations (RTSs). (**b**) RTSs are clustering into scenarios based on their timing and energy costs. (**c**) The monitoring parameters and the system costs of scenarios and the monitoring parameters are recorded in a table. (**d**) At run-time, system costs are predicted based on the scenarios

deterministic behaviors the system can encounter at run-time, such as the body of a loop, a certain function the system needs to execute, or the read access to a certain chunk of data. In addition, each RTS is associated with its system costs, such as execution time and energy consumption. With different settings of knobs, the system costs of an RTS can vary and form a trade-off curve. As shown in Fig. 5.9b, each curve in the time-energy plane represents the system costs of an RTS.

Since the number of RTSs is too many to be recorded for run-time usage at low costs, the RTSs are further clustered into scenarios. The RTSs in the same scenario share some similar properties which become the identity of the scenario. In general, the RTSs in the same scenario are required to be similar in the system cost trade-offs in executing them under different knob settings, so that the system cost of each RTS will be approximated by the aggregated values (average,

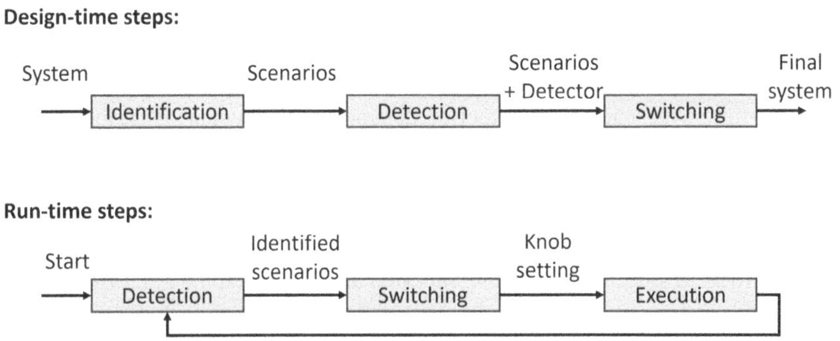

Fig. 5.10 Basic flow of realizing a solution to dynamism-related problems based on dynamic scenario principles

maximum, minimum, etc.) of the system costs of all RTSs in the same scenario. Figure 5.9b visualizes this process, in which the clustering is equivalently the segmentation of the multidimensional cost space into regions. After clustering, each scenario contains multiple RTSs with similar costs. Then, *monitoring parameters* are discovered, which are the parameters used to distinguish scenarios at run-time. The monitoring parameters and the system costs of each scenario are summarized and recorded in a table like Fig. 5.9c.

At run-time, based on the reference of the scenarios, the controller can identify the scenarios of the current and upcoming RTSs and then predict the future system costs with limited errors. Based on accurate system cost prediction, the controller can derive the optimal configuration of the system knobs for the upcoming workload. In addition, the costs can be updated and the scenarios can be re-clustered based on the monitoring at run-time, which further enhances the adaptability of dynamic scenarios to variability.

The concept of dynamic scenarios leads to a generic flow consisting of several design-time and run-time steps for realizing a solution to any dynamism-related problem in different contexts. The basic flow of these steps is shown in Fig. 5.10. At design time, the required steps are described as follows:

1. **Identification of the scenario set:** In this step, the solution space of RTS scenario mapping is explored, and an appropriate scenario set is determined. At the beginning of this step, the system needs to be profiled with high coverage to obtain the system costs of every possible situation of the system. Based on the profiling results, the RTSs are defined. The profiling approach and RTS definition are dependent on the system. Then, the profiled system costs of every RTS become the reference of the following scenario set creation.
2. **Detection of the scenario:** In this step, a function that maps an RTS to the scenario it belongs to is created, called the scenario detector. In practice, the system will execute this function at run-time to identify the scenario of the encountered RTSs. Normally, the detector is realized in the form of a look-up

table, in which the indices are the *monitoring parameters* . Normally, monitoring parameters can be certain system parameters and application variables which influence a lot on the system costs, such as the input size of the application. Since the run-time overhead of the detection depends on the scenario set, the scenario set may be refined in this step to reduce the detection overhead.
3. **Switching from one scenario to another:** Switching is the act of changing the system from one knob setting to another. Since the system cost of RTSs varies with the knob setting, different knob settings can be preferred for the same RTS under different system conditions. This step aims to develop an algorithm which is used at run-time to decide whether to switch the knob setting and how to switch the knob setting, based on the scenarios detected at run-time. Since switching incurs run-time overhead, it is not always preferred to switch to the optimal knob setting for the scenario. Therefore, the algorithm should consider both the system costs of the scenario and the switching overhead at the same time in making switching decisions.

The final products of the design-time steps are the scenario set, the scenario detector, and the algorithm for the switching decisions. These products are stored and used at run-time for predicting the dynamism and on-the-fly optimization of the system. A run-time controller is implemented which uses the scenario set to run the scenario detector and the algorithm to switch the knob setting of the system. In particular, the basic run-time steps performed in order by the controller are described as follows:

1. **Detection of the scenario:** In this step, the scenario detector is called to identify the scenarios of the RTSs the system encounters now and will encounter in the near future. The controller reads the monitoring parameters and uses the function defined in the scenario detector to acquire the scenarios of the encountered RTSs. With the scenario information, the bound of the system costs of the encountered RTSs can be known in advance by the controller.
2. **Switching from one scenario to another:** In this step, the algorithm is called to decide on whether to switch the knobs and how to switch the knobs, according to currently known scenarios of the future RTSs. Based on the result of the algorithm, the knob setting is switched or unswitched accordingly.
3. **Execution of the scenario:** In this step, the system executes the current RTS under the knob setting decided in the last step. Then the controller returns to the first step for the following RTSs.

These steps are repeated continuously so that the controller deals with the encountered RTSs in sequence and makes rolling decisions as the system proceeds.

In the context of performance variability mitigation, the above dynamic scenario principles are instantiated to develop a scheduling algorithm that controls multiple knobs to handle performance variability for timing guarantees. Each of the aforementioned steps is implemented in the instantiated methodology but tailored for the specific purpose of performance variability mitigation. In the following sections, the details of the methodology are elaborated.

5.4.2 *Design-Time Scenario Clustering*

At design time, the scenario set, the scenario detector, and the controlling algorithm are generated. The scenario set and the scenario detector should be tailor-made for the application context. Therefore, the detailed implementation of design-time steps can be different from one application to another. In this section, we only describe the general process of design-time steps without addressing any application-specific aspects, while Sects. 6.2 and 6.3 show two examples of design-time steps of creating the application-specific scenario set and the scenario detector.

Figure 5.11 shows the process of design-time steps for creating the application-specific scenario set. Since scenarios are grouped mainly based on system costs, it requires complete data on the system costs in executing each part of the application. Therefore, the application first needs to be thoroughly profiled in a finer timescale than the time granularity of the mitigation. In our approach, RTL-level simulation is performed to profile the application, which is accurate down to the cycle level, approximately the nanosecond level, much finer than the sub-millisecond time granularity of the mitigation. The application also needs to be profiled with different inputs and configurations to traverse as many different execution paths in the data flow as possible, so that the variability in the system costs of executing the application is thoroughly explored. The existing techniques for improving profiling coverage can be utilized, such as data flow analysis [3].

Then, the process of scenario set generation is performed, depicted in Fig. 5.11. First, the profiling traces of the workload are segmented into atomic units called *thread node* (TN). The TNs are further categorized into RTSs, in which TNs with identical roles in the data flow or identical behavioral characteristics are labeled as the same RTS. For example, TNs that perform different iterations of the same loop can have equivalent roles in the data flow and identical system costs, so they can be categorized into the same RTS. Essentially, a workload can be seen as a sequence of TNs, each of which instantiates the same or different RTSs.

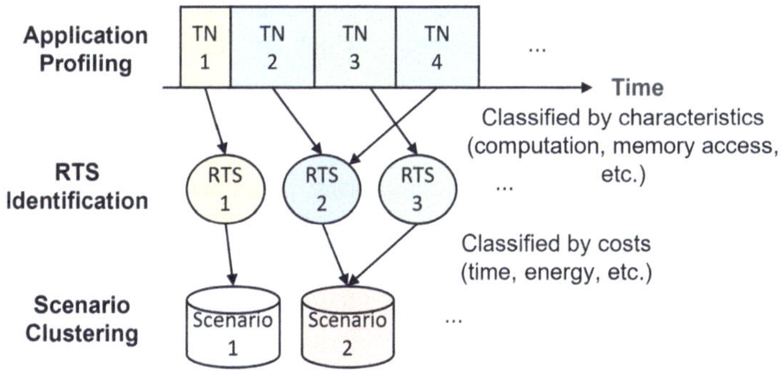

Fig. 5.11 Process of generating a scenario set in the design-time process of the dynamic scenario-based approach

There are no strict definitions of the size and the content of a TN or an RTS. However, since TNs and RTSs are the smallest executable units used in the scheduling of the control algorithm, their execution time should match the time granularity of the mitigation. In our approach, HDP enables mitigation down to the job size of the order of tens of microseconds, so each TN and RTS has an execution time of approximately tens of microseconds.

TNs and RTSs can be defined either manually or automatically. The manual method relies on the knowledge of the software engineers on the data flow to group the segments of the profiling traces into RTSs. This method is applicable if the data flow is completely known by the engineers. However, for complicated applications, their data flows can be too difficult for the engineers to analyze. Therefore, automatic data flow analysis algorithms can be utilized to systematically analyze the data flow of the application and identify RTSs [3].

Each RTS has its system costs (execution time, energy consumption), which also depends on the knob setting that is used when executing the RTS. Therefore, the system costs of each RTS can be represented by a trade-off curve of multiple cost factors, in which each point on the curve represents the system cost of the RTS under a certain knob setting.

Thereafter, the RTSs are further clustered into dynamic scenarios. The primary consideration of the scenario clustering is the similarity in system costs. At run-time, the system cost of each RTS will be approximated by the aggregated values (average, maximum, minimum, etc.) of the system costs of all RTSs in the same scenario. The error of this approximation will only be small if the differences in the system costs of RTSs in the same scenario are minimal.

During scenario clustering, the scenario detector is created at the same time, which can map an RTS to its belonging scenario based on the monitoring parameters. At run-time, the controller should be able to run the scenario detector to identify the scenario of an RTS by the monitoring parameters, with low overhead of system costs. The choices of monitoring parameters are another important consideration in scenario clustering. To reduce the run-time overhead, the monitoring parameters should be easily read at run-time, such as application variables or metadata of the input data. The monitoring parameters should also have high impact on the system costs of the RTSs, so the scenarios, which are clustered by the monitoring parameters, can be distinct in system costs from one to another.

A common form of the scenario detector is a look-up table in which the indices are the monitoring parameters, and the look-up values are the belonging scenario of the RTS with the corresponding monitoring parameters. Hence, at run-time, detecting the scenario of an RTS is done by reading the monitoring parameters of the RTS and using them as indices to look up the table. In this form of the scenario detector, each scenario can be seen as a region in the multidimensional space of the monitoring parameters. Therefore, the problem of grouping the RTSs into scenarios can be modeled as the segmentation of the parameter space into regions.

In practice, the discovery of monitoring parameters and the scenario clustering can be performed manually by analyzing the execution flows, if the complexity of the application is not too high. For applications of high complexity or unknown execution flows, automatic clustering algorithms can be developed on the same principles to perform these tasks systematically [6].

After clustering, statistics (worst RTS cost, average RTS cost, etc.) for each scenario are stored in a look-up table to be used at run-time for the calculation of execution time. In this approach, we record the worst execution cycles and the average execution cycles for the RTSs of each scenario. At run-time, the worst execution cycles are used as the upper bounds of execution cycles for ensuring timing guarantees. On the other hand, the average execution cycles are used as the references for the prediction of execution cycles for normal cases. The monitoring parameters are also recorded as the indices of the look-up table. At run-time, detecting a scenario is merely querying the look-up table by the monitoring parameters.

In practice, the scenario table can be stored in a specific region of the program codes of the application, which can be read by the controller at run-time. Alternatively, the table can be stored in a specific region in the memory of the system, if the system is domain-specific that only runs a limited number of applications.

5.4.3 Principles of the Run-Time Scheduling Algorithm

At run-time, the exact execution time of the workload depends on the actual inputs, such as the compressed image to be decoded by the image decoder, or the sound waves to be encoded by an audio encoder. By applying different inputs, the variability in execution time is injected into the system, which could potentially cause deadline misses. Nevertheless, the algorithm has access to the dynamic scenario table, which contains parameterized bound and expected execution time based on the correlation-based analysis performed at design time, elaborated in the last section. With these references, the run-time controller can obtain the upper bound of the execution time with full guarantee, as well as predict the execution time with limited discrepancy. Therefore, with proper scheduling of the modes of the knobs, variability in execution time can be offset before the deadlines, so full timing guarantees can be ensured.

The principles of generic DS-instantiated scheduling algorithms have been developed and elaborated in [9], in which the essential steps of the algorithm are defined and shown in Fig. 5.12. Here, we summarize the principles from step to step as follows:

- **Look ahead subsequent TNs:** The controller gathers the updated information of the subsequent TNs, so that the current TN and subsequent TNs are considered together in the following decision process. The current TN and the subsequent TNs that the controller can look ahead form the optimization window of the mode scheduling of the knobs. In this way, the controller can avoid selecting knob settings that are only optimal for the current TN but suboptimal in the long run. With dynamic scenarios, the action taken in this step is actually to identify the

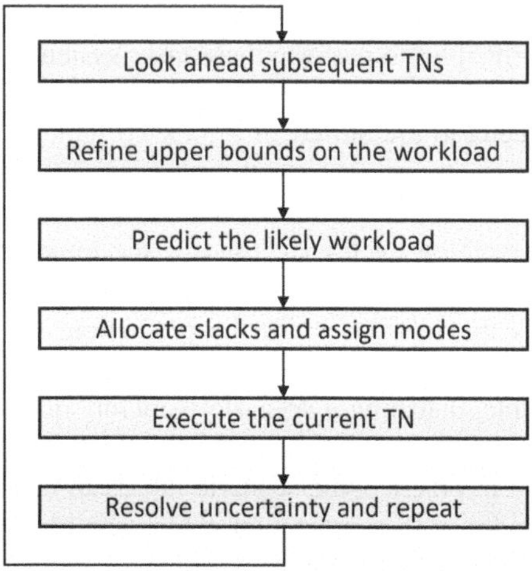

Fig. 5.12 The generic flow of DS-instantiated scheduling algorithms

scenarios of these TNs and look up the scenario-specific information brought from the design time.

- **Refine upper bounds on the workload:** Given the information obtained in the last step, the controller calculates the refined upper bounds of the uncertainty in the execution time of the TNs in the optimization window. The bounds are used to derive the maximum slacks that can be exploited while ensuring timing guarantees. The tighter the bounds, the larger the slacks. Thus, the bounds should be as tight as possible to maximize the slacks.
- **Predict the likely workload:** The controller predicts the most likely execution time of the TNs in the optimization window. The predictions are used in the following steps to figure out the knob settings that are most likely to be optimal over the optimization window, but not for the extreme cases that rarely happen.
- **Allocate slacks and assign modes:** The controller solves the optimal slack allocation and mode assignment over all TNs in the optimization window. The amounts of slacks need to be calculated by the worst case of ensuring timing guarantees, in which the hardest deadline constraints and the upper bound execution time of each TN are assumed. Only when the slack is available beyond what is required to meet the hardest deadline, then that slack can be used on the TNs. Once the amount of slacks is obtained, the optimal slack allocation and mode assignment should be solved with the predicted most likely execution time of the TNs. The switching overheads also need to be considered in this optimization problem, so the slack should be particularly allocated for the switching time and the switching energy should add to the total energy cost, which makes frequency switching undesirable. Due to the switching overheads and the discrete modes of knobs, the derivation of slack allocations and knob settings becomes a nonlinear optimization problem, which could not be solved with small computation effort. Hence, heuristics need to be applied to find near-optimal solutions in practice.

- **Execute the current TN:** The controller switches the knob to the assigned mode for the current TN. Then, the TN is executed.
- **Resolve uncertainty and repeat:** Once the current TN completes its execution, its uncertainty is resolved and its exact execution time is known. Also, more information about the subsequent TNs can become available during the execution time. Thus, the controller needs to revise its optimal schedule with the new and updated information by repeating the previous optimization process. In this way, the controller keeps incrementally optimizing the schedule as the uncertainties gradually resolve over time, so near-optimal decisions are reached at any time during the workload.

5.4.4 Run-Time Scheduling for a Single Knob

Based on the principles described in the last section, we have implemented the DS-instantiated scheduling algorithm which supports both single-knob and multiple-knob settings. In the following, we describe the single-knob algorithm, showing the essential procedure of the algorithm. The algorithm assumes a knob with negligible switching overheads, such as HDP. Then, in the next section, we further describe the full multiple-knob algorithm, showing how the algorithm is extended to handle the switching overheads and assign modes of multiple knobs.

The overall procedure of the run-time single-knob scheduling is shown in Algorithm 5.1. At the start of the workload, an *optimal schedule* is generated by scheduling with the average number of cycles of each TN and the hard deadlines of the workload. This schedule is the presumed ideal schedule which can ensure timing guarantees and minimize energy consumption if there is no variability. The goal of the algorithm is to keep the progress ahead of the optimal schedule, but only by a necessary amount of slack required to ensure timing guarantees. Then, the controller defines a deadline for each TN, which is equal to the TN finish time in the optimal schedule. These deadlines serve as checkpoints for the progress of the workload. In addition, the controller creates a buffer to keep track of the scenarios of the TNs in the optimization window, so the controller can look into the buffer to evaluate the future workload.

At the start of the execution of each TN, the controller calls the scenario detector to identify the scenario of the last TN in the buffer [8]. Then, two processes are performed to decide the mode of the knob to run the next TN. First, the controller calculates the timing guarantee criteria by a *refined upper bound* procedure, shown in Algorithm 5.2. The timing guarantee criteria are expressed by the minimum required speed for the next TN, which is the slowest speed to run the next TN while still ensuring timing guarantees. Figure 5.13 gives a visual illustration of the principle of this procedure. To calculate the minimum required speed, the controller generates a schedule which could create the greatest possible slack. This means that the controller assumes the other scenarios in the buffer to be their worst execution time and run at the fastest mode. If gas pedals mentioned in Sect. 3.2.3 exist, the

Algorithm 5.1 Run-time scheduling

Input: thread nodes $TN_{1...N}$, buffer size B
Output: modes which each TN uses $m_{1...N}$
$CB_{1...N} \leftarrow$ cycle budgets of $TN_{1...N}$
$DL_{i...N} \leftarrow OptimalSchedule(CB_{1...N})$
Initialize an empty buffer buf
for $i \leftarrow 1$ to B **do**
    $S_i \leftarrow ScenarioDetect(TN_i)$
    Push S_i to buf
end for
for $i \leftarrow 1$ to N **do**
    $f_{mr} \leftarrow RefinedUpperBound(buf)$
    $f_{opt} \leftarrow LikelyPrediction(buf)$
    $m_i \leftarrow$ the nearest faster mode of $max(f_{mr}, f_{opt})$
    Execute TN_i under mode m_i
    Pop S_i from buf
    if $i + B \leq N$ **then**
        $S_{i+B} \leftarrow ScenarioDetect(TN_{i+B})$
        Push S_{i+B} to buf
    end if
end for

Algorithm 5.2 Refined upper bound

procedure REFINEDUPPERBOUND(buf)
    $T_{cur} \leftarrow$ current time
    $f_{max} \leftarrow$ maximum frequency
    $B \leftarrow$ number of TNs in buf
    $DL_{1...B} \leftarrow$ deadlines of TNs in buf
    $WC_{1...B} \leftarrow$ worst cycles of scenarios of TNs in buf
    $PDL_B \leftarrow DL_B$
    for $i \leftarrow B$ **downto** 1 **do**
        $PDL_{i-1} \leftarrow min(PDL_i - WC_i/f_{max}, DL_{i-1})$
    end for
    $f_{mr} \leftarrow WC_1/(PDL_1 - T_{cur})$
    return f_{mr}
end procedure

controller also needs to assume that the gas pedal is activated to run for its maximum allowable time and then switch to the normal modes. This procedure is performed for each hard deadline of the TNs in the buffer, and the hardest speed constraint of all is considered in the end.

Second, *likely prediction* is performed to calculate the optimal processor speed that is most likely to minimize the total energy consumption for the TNs in the buffer. This process is shown in Algorithm 5.3. At the beginning of the procedure, it would predict the real execution time of the TNs in the buffer, based on the statistics of the scenarios acquired in the design-time process. Heuristics can be applied to improve the accuracy of the prediction. In this work, we predict the TN execution time to be the average execution time of the scenario, which is measured during the

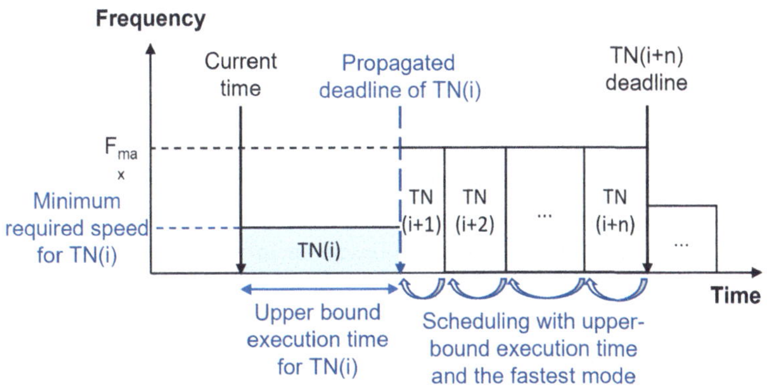

Fig. 5.13 Refined upper bound process to calculate the minimum required speed for the next TN for ensuring timing guarantee

Algorithm 5.3 Likely prediction

procedure LIKELYPREDICTION(buf)
  $T_{cur} \leftarrow$ current time
  $B \leftarrow$ number of TNs in buf
  $DL_{1...B} \leftarrow$ deadlines of TNs in buf
  $AC_{1...B} \leftarrow$ average cycles of scenarios of TNs in buf
  $f_{opt} \leftarrow sum(AC_{1...B})/(DL_B - T_{cur})$
  return f_{opt}
end procedure

design-time scenario clustering, as described in Sect. 5.4.2. Based on the execution time prediction, the scheduler generates an optimal schedule that minimizes the total energy consumption for the TNs in the buffer, and thus the mode to run the next TN is decided. Solving the optimal schedule at run-time will not be feasible since it takes too much computation, so heuristics need to be applied. In this work, our heuristics are designed based on the concept that it is near-optimal to run constantly at the slowest possible speed. Therefore, the controller calculates the optimal speed which is just enough to run all TNs to meet the deadline of the last TN in the buffer. Finally, the controller finds the slowest mode that is faster than both the optimal speed and the minimum required speed, as shown in Line 12 of Algorithm 5.1. This becomes the selected mode to run the next TN.

The algorithm above can be easily extended to other contexts and compatible with other scheduling techniques. For example, for scheduling for multiple applications, real-time scheduling [2, 11] can be performed on top of the algorithm, deciding the order of running applications. Then, our algorithm handles fine-grained time resource arrangement and performance variability management. In addition, the maximal latency specification of general applications can be treated as a single deadline. For time-sensitive applications, providing a strict bound on the maximal deadline misses can be realized by statistical analysis at design time and ignoring a number of the hardest deadlines at run-time.

Still, there is room for improvement in the algorithm. In this approach, the detection of scenarios at run-time is purely based on the monitoring parameters defined at design time, which is viable for simple workloads, such as periodic frame-based applications we use in the experiments later in Chap. 6. However, scenario detection is generally predicting future workload, which is not an easy task in realistic scenarios with complicated interactions among multiple applications and software abstraction layers. In future work, scenario clustering can be potentially generalized and enhanced by predictors based on inference of machine learning models, so it can handle any application context. Moreover, switching scenarios as often as every TN might not be optimal due to switching overheads, so it is crucial to decide when to switch from one scenario to a new one. The conceptual work on switching decisions has been developed in [8], which can be applied to improve the current approach in the future.

5.4.5 Run-Time Scheduling for Multiple Knobs

To exploit the advantages of both HDP and DVFS, the run-time scheduling is adapted for multiple knobs. The flow diagram of the enhanced run-time procedure is shown in Fig. 5.14. In principle, the algorithm sequentially decides the modes of the knobs from the slow-switching ones to the fast-switching ones. In our approach, the DVFS mode is decided first, and then the HDP mode is decided. Moreover, the switching rates of HDP and DVFS are different. HDP can switch at the start of each TN. On the other hand, DVFS can only switch once over multiple TNs. Otherwise, its switching time overhead would greatly lengthen the execution time.

Here, we limit the interval between two DVFS switches to be larger than a threshold value $T_{th,DVFS}$. Specifically, the controller would consider switching the DVFS knob only when the time since the last time the controller decided the DVFS mode is more than $T_{th,DVFS}$. The choice of $T_{th,DVFS}$ is a balance between the granularity of mitigation and the overhead on execution time. If $T_{th,DVFS}$ is too long, the DVFS might not be able to react to the variability faster than $T_{th,DVFS}$. On the other hand, if $T_{th,DVFS}$ is too short, the DVFS might switch too frequently so that the switching time significantly lengthens the execution time, as illustrated in the example of Fig. 5.1 in Sect. 5.1. In our approach, we set $T_{th,DVFS}$ to 10 ms, so that the DVFS switching time overhead is less than 1% of the execution time. The principle of our approach can be extended to any knobs of any timescales.

In principle, the procedure of deciding modes for multiple knobs is compatible with any algorithm. Nevertheless, the DS-based algorithm can calculate a tight upper bound of execution time, so it can exploit the benefits of multiple knobs much better than other algorithms. Algorithm 5.4 shows the detailed flow of the run-time procedure of the DS-based algorithm for multiple knobs. In Line 10, the controller would check whether the time since the last DVFS decision point is longer than $T_{th,DVFS}$ to decide whether to decide a new DVFS mode.

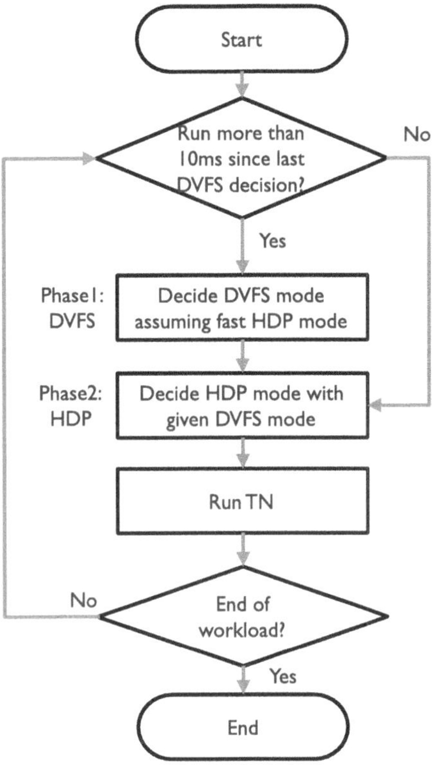

Fig. 5.14 Flow diagram of the run-time process of the multi-timescale DS based approach, in which the modes of knobs are decided sequentially from the slow-switching ones to the fast-switching ones

In Line 11, the controller calls the enhanced refined upper bound procedure . The detail of the procedure is shown in Algorithm 5.5. A major change is that the controller would consider that the DVFS mode must run over multiple TNs until the next DVFS switch, as illustrated in Fig. 5.15. Therefore, to generate the case of the most positive slack, the processor is assumed to run in a certain DVFS mode for $T_{th,DVFS}$ and then switch to the fastest DVFS mode until the end of the buffer. The minimum speed requirement (f_{mr}) is derived based on this case, as shown in Lines 15–19. In addition, the controller also derives the speed requirements to meet all deadlines before the next DVFS switch. Based on all of these requirements, the controller derives the lowest speed which can ensure all timing guarantees, as shown in Lines 22–24.

In Line 12, the controller calls the enhanced likely prediction procedure , which considers DVFS switching time in calculation. The detail of the process is shown in Algorithm 5.6. Finally, in Line 14–18, the controller would only decide the new DVFS mode when DVFS is triggered. Otherwise, the current DVFS mode will be used for the next TN. To ensure the timing guarantees, when deciding on the new DVFS mode, the controller assumes the fastest HDP mode is used. Once the DVFS mode is decided, the new HDP mode will be decided based on the condition of the selected DVFS mode.

Algorithm 5.4 Run-time scheduling for multiple knobs

Input: thread nodes $TN_{1...N}$, buffer size B
Output: DVFS & HDP modes which each TN uses $m_{DVFS,1...N}, m_{HDP,1...N}$
$CB_{1...N} \leftarrow$ cycle budgets of $TN_{1...N}$
$DL_{i...N} \leftarrow OptimalSchedule(CB_{1...N})$
Initialize an empty buffer buf
for $i \leftarrow 1$ to B **do**
  $S_i \leftarrow ScenarioDetect(TN_i)$
  Push S_i to buf
end for
for $i \leftarrow 1$ to N **do**
  $DVFS\_triggered \leftarrow CheckDvfsTriggered()$
  $f_{mr} \leftarrow RefinedUpperBound\_MT(buf)$
  $f_{opt} \leftarrow LikelyPrediction\_MT(buf, DVFS\_triggered)$
  $f_{res} \leftarrow max(f_{mr}, f_{opt})$
  if $DVFS\_triggered$ **or** $i = 1$ **then**
    $m_{DVFS,i} \leftarrow$ the nearest faster DVFS mode of f_{res} assuming using fastest HDP mode
  else
    $m_{DVFS,i} \leftarrow m_{DVFS,i-1}$
  end if
  $m_{HDP,i} \leftarrow$ the nearest faster HDP mode of f_{res} assuming using $m_{DVFS,i}$
  Execute TN_i under mode $m_{DVFS,i}, m_{HDP,i}$
  Pop S_i from buf
  if $i + B \leq N$ **then**
    $S_{i+B} \leftarrow ScenarioDetect(TN_{i+B})$
    Push S_{i+B} to buf
  end if
end for

5.4.6 *Overheads of the Algorithm*

Running the algorithm could induce execution time overhead. The amount of time overhead depends on the hardware implementation of the system. If the algorithm runs in parallel with the original workload, then most of the execution time overhead can be hidden. Normally this means the algorithm is run in a distinct hardware, such as the power management unit or a dedicated processor core.

If the algorithm and the original workload share the same hardware resource (e.g., they run in the same processor core), then the algorithm costs extra execution time at the beginning of each TN of the original workload. The influence of this time overhead is similar to how the switching time of the knobs affects the time overhead and the time granularity of the mitigation, as discussed in Sect. 3.2. The total time overhead of the approach for each TN is actually the switching time of the knobs plus the execution time of the algorithm. In order to minimize the performance degradation and improve the time granularity of the mitigation, both the switching time and the execution time of the algorithm should be small compared to the original workload. Since our approach operates at the time granularity of the deca-

Algorithm 5.5 Refined upper bound for multiple knobs

procedure REFINEDUPPERBOUND_MT(buf)
  $T_{cur} \leftarrow$ current time
  $f_{max} \leftarrow$ maximum frequency
  $N \leftarrow$ number of TNs in buf
  $DL_{1...B} \leftarrow$ deadlines of TNs in buf
  $WC_{1...B} \leftarrow$ worst cycles of scenarios of TNs in buf
  $T_{sw,DVFS} \leftarrow$ DVFS switching time
  $T_{th,DVFS} \leftarrow$ period between two DVFS switching

  // Propagate deadlines
  $PDL_B \leftarrow DL_B$
  for $i \leftarrow B$ **downto** 1 **do**
    $PDL_{i-1} \leftarrow min(PDL_i - WC_i/f_{max}, DL_{i-1})$
  end for

  // Find the last TN of DVFS interval
  $j \leftarrow 1$
  while $PDL_j < T_{cur} + T_{th,DVFS} + 2T_{sw,DVFS}$ **do**
    $j \leftarrow j + 1$
  end while
  $f_{mr} \leftarrow sum(WC_{1...j})/(PDL_j - T_{cur} - 2T_{sw,DVFS})$

  // Check deadlines in this DVFS interval
  for $k \leftarrow 1$ to j **do**
    $f_{mr} \leftarrow max(f_{mr}, sum(WC_{1...k})/(DL_k - T_{cur} - T_{sw,DVFS}))$
  end for
  return f_{mr}
end procedure

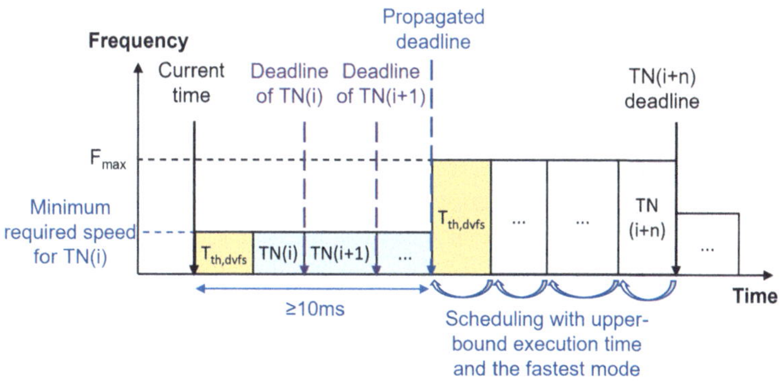

Fig. 5.15 Enhanced refined upper bound process to calculate the minimum required speed for the next TN for ensuring timing guarantee in the multiple-knob scheduling algorithm

microsecond level, the execution time generating a decision in the algorithm should be at most hundreds of nanoseconds, approximately a thousand clock cycles for a normal processor of several GHz clock frequency.

Algorithm 5.6 Likely prediction for multiple knobs

procedure LIKELYPREDICTION_MT(buf, $DVFS\_triggered$)
  $T_{cur} \leftarrow$ current time
  $B \leftarrow$ number of TNs in buf
  $DL_{1...B} \leftarrow$ deadlines of TNs in buf
  $AC_{1...B} \leftarrow$ average cycles of scenarios of TNs in buf
  $T_{sw,DVFS} \leftarrow$ DVFS switching time
  if $DVFS\_triggered$ **then**
    $f_{opt} \leftarrow sum(AC_{1...B})/(DL_B - T_{cur} - T_{sw,DVFS})$
  else
    $f_{opt} \leftarrow sum(AC_{1...B})/(DL_B - T_{cur})$
  end if
  return f_{opt}
end procedure

Therefore, the algorithm design is restricted by the maximum execution time requirement. The scenario detector and the buffer size are the two main factors which affect the execution time of the algorithm. Although a complicated scenario set and scenario detector can reduce the errors of system cost prediction and bounding, they increase the execution time overhead. Also, a larger buffer size considers more deadline constraints and thus requires more execution time. Hence, the complexity of the scenario detector and the size of the buffer should be limited to meet the execution time constraint. A typical implementation uses several scenarios, a scenario detector based on a look-up table, and a buffer size of at most 20, which are lightweight that the execution time is negligible compared to the execution time of a TN.

5.4.7 Comparison of System, Adaptive, and Dynamic Scenarios

Figure 5.16 underlines the difference between the dynamic, the adaptive, and the system scenarios concept. In system scenarios, the scenario clustering takes place offline and at run-time, a scenario detector matches the current RTS to each related scenario, while the controller decides the knob setting. In the adaptive scenarios, a run-time, scenario re-clustering procedure occurs which is based on the monitored RTS history. Finally, in the dynamic scenarios approach, the workload predictor allows us to foresee future RTSs and apply proactive mitigation.

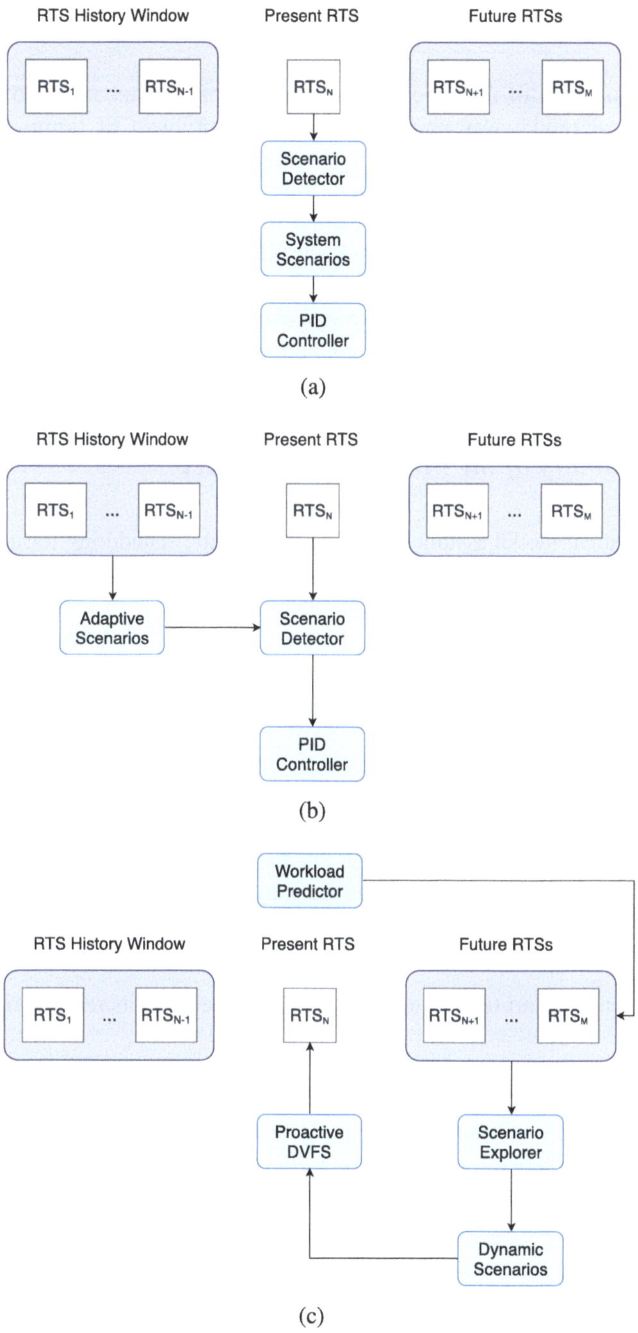

Fig. 5.16 An illustration of the System, Adaptive, and Dynamic Scenarios methodology: (**a**) System Scenarios, (**b**) Adaptive Scenarios, and (**c**) Dynamic Scenarios

5.5 Proof of Timing Guarantees

In the previous work of the conceptual framework of dynamic scenario principles, a mathematical model was presented, where the problem of timing-guaranteed scheduling was formulated as a set of optimization functions and constraints [9]. Our proposed algorithm is instantiated from the conceptual framework, in the context of sub-millisecond-level mitigation of highly dynamic workloads, on the processor equipped with speed-varying knobs like HDP and DVFS. In the following, we will show how the algorithm is transferred from the mathematical model, while the hard constraints are fulfilled in the proposed algorithm. Hence, we prove that the algorithm intrinsically ensures timing guarantees in our specific context.

5.5.1 Summary of the Mathematical Model

The mathematical model assumes typical deterministic scheduling for a number of consecutive jobs with uncertain execution cycles and hard deadlines, executing on a generalized processor model of several operating modes with distinct cycle times. In a nutshell, the processor is assumed to operate in cycles, while the cycle time $CT[m]$ varies with its operating mode m. Switching from a mode m_a and another mode m_b takes switching time of at most $O(m_a, m_b)$. The processor is required to execute a sequence of jobs $J_1 \cdots J_N$, each of which has its own deadline D_i, ready time R_i, upper bound execution cycles UC_i, and likely execution cycles LC_i. The objective of the scheduler is to find a proper start time s_i and an operating mode m_i for each job, which can ensure that all jobs finish before their deadlines even though the exact execution cycles of each job are unknown before execution. Table 5.2 enumerates the detailed denotations of the parameters in this model. Based on the model, the hard constraints required for a static schedule to ensure timing guarantees are derived as follows:

Constraint 1: The start time of a job cannot be earlier than its ready/arrival time:

$$s_i \geq R_i \tag{5.1}$$

Constraint 2: The start time of a job cannot be earlier than the expected completion time of the previous job plus switching time overhead:

$$s_{i+1} \geq s_i + LC_i \cdot CT[m_i] + O(m_i, m_{i+1}) \quad \forall i \geq 1 \tag{5.2}$$

Boundary case of the above constraint for the first job:

$$s_1 \geq PCT + O(m_0, m_1) \tag{5.3}$$

Table 5.2 Denotations of the mathematical model of scheduling with hard timing constraints

Processor characteristics	
M	Number of distinct operating modes provided by the processor to execute a job
$CT[1 \ldots M]$	A constant array where $CT[m]$ is the effective cycle time of the processor when in mode m
$O(m_a, m_b)$	Worst-case time overhead in switching from mode m_a to mode m_b
F	Fastest mode; independent of job and processor state
m_0	Mode setting for the job prior to the first job J_1
Job characteristics	
N	Number of consecutive jobs considered in the joint optimization by the scheduler at a given scheduling instance
R_i	Ready/arrival time of job J_i $\forall 1 \leq i \leq N$; execution of a job cannot be started before this time even if the processor is free
D_i	Hard deadline of job J_i; execution of a job must be completed by this time even when the job has its worst-case workload; each job can have its own deadline
UC_i	Refined upper bound execution cycles of J_i; scheduler invokes upper bound refinement model at run-time for each job to get this info
LC_i	Predicted likely execution cycles of J_i (given by the estimator)
$LC_i \cdot CT[m]$	Likely duration of J_i if executed with mode m
$UC_i \cdot CT[m]$	Guaranteed maximum amount of time J_i can take if executed with mode m
LS_i	Latest (or upper bound on) start time of J_i, i.e., if the job J_i is not started by this time one or more of deadlines $D_k, k \geq i$ might miss which is not acceptable
W	Safe mode of the job subsequent to J_N; if the processor is in mode W by the time D_N, schedulability of subsequent jobs beyond the optimization horizon is guaranteed; $W \leq F$ assuming mode with lower index is faster mode
PCT	Completion time of the job prior to J_1; namely current time of the processor
Variables to be solved by the scheduler	
$s_1 \ldots s_N$	Likely start time of job $J_1 \ldots J_N$
$m_1 \ldots m_N$	Likely mode setting for job $J_1 \ldots J_N$

Constraint 3: A job must complete its execution before the deadline even for the upper bound workload:

$$s_i + UC_i \cdot CT[m_i] \leq D_i \quad \forall i < N \tag{5.4}$$

Boundary case of the above constraint for the last job:

$$s_N + UC_N \cdot CT[m_N] \leq D_N - O(m_N, W) \tag{5.5}$$

Constraint 4: Even when the current and future jobs pose upper bound workload, there should be enough time either (i) to switch to the fastest mode and execute all the later jobs using the fastest mode and complete them before their respective deadlines or (ii) selected modes for all jobs are such that all deadlines will be met even for the worst-case workload conditions:

$$s_i + UC_i \cdot CT[m_i] \leq \max\{ s_{i+1} - O(m_i, m_{i+1}), \\ LS_{i+1} - O(m_i, F)\} \quad \forall i < N \tag{5.6}$$

where

$$LS_N = D_N - UC_N \cdot CT[F] \tag{5.7}$$

$$LS_i = \min\{D_i, LS_{i+1}\} - UC_i \cdot CT[F] \quad \forall i < N \tag{5.8}$$

5.5.2 Instantiation of the Mathematical Model

On the foundation of the conceptual framework, we develop the proposed algorithm for sub-millisecond-level scheduling in a processor core with DVFS and HDP. Therefore, the variables of the model are mapped into the variables in the algorithm, representing the properties of the real processors and applications. Table 5.3 lists all the relations between the variables of the model and those of the algorithm. In addition, we will elaborate on how the mapping is made as follows:

- Each operating mode m represents a pair of a DVFS mode and a HDP mode (m_{DVFS}, m_{HDP}). Therefore, $CT[m]$ corresponds to the clock period when the processor is running under this combination of DVFS mode and HDP mode, which can be denoted as $CT[m_{DVFS}, m_{HDP}]$
- $O(m_a, m_b)$ is dependent on whether DVFS mode switching or HDP mode switching is involved. Assuming the worst DVFS switching time is $T_{sw,DVFS}$ and the worst HDP switching time is $T_{sw,HDP}$, then $O(m_a, m_b)$ can be denoted as follows:

$$O(m_a, m_b) = \begin{cases} 0 & \text{if } m_{DVFS,a} = m_{DVFS,b} \\ & \text{and } m_{HDP,a} = m_{HDP,b} \\ T_{sw,DVFS} & \text{if } m_{DVFS,a} \neq m_{DVFS,b} \\ & \text{and } m_{HDP,a} = m_{HDP,b} \\ T_{sw,HDP} & \text{if } m_{DVFS,a} = m_{DVFS,b} \\ & \text{and } m_{HDP,a} \neq m_{HDP,b} \\ T_{sw,DVFS} + T_{sw,HDP} & \text{otherwise} \end{cases} \tag{5.9}$$

Table 5.3 Relations between the variables in the math model and in our proposed algorithm

Variable in the model	Related variable in the algorithm	Note
Processor characteristics		
m	(m_{DVFS}, m_{HDP})	Each operating mode represents a pair of a DVFS mode and an HDP mode
$CT[m]$	$CT[m_{DVFS}, m_{HDP}]$	$CT[m]$ is a function of the DVFS mode and the HDP mode
$O(m_a, m_b)$	$T_{sw,DVFS}$, $T_{sw,HDP}$	$O(m_a, m_b)$ is related to the switching time of DVFS and HDP, defined by Eq. 5.9
F	$f_{\max}$	$f_{\max}$ denotes the fastest clock frequency of all modes, equal to $1/CT[F]$
m_0	(n/a)	m_0 is considered to be unknown
Job characteristics		
N	B	The number of jobs to be scheduled is the same as the buffer size B
J_i	TN_i	Each job represents a thread node
R_i	(n/a)	The constraint of the ready time is always fulfilled. Hence, R_i is not required in the algorithm
D_i	DL_i	Hard deadline of TN_i
UC_i	WC_i	UC of each TN is the worst-case execution cycles of the scenario
LC_i	AC_i	LC of each TN is the average execution cycles of the scenario
LS_i	PDL_i	LS_i and PDL_i, respectively, denote the latest start time and the latest end time for J_i. Their relation can be expressed by Eq. 5.15
W	(n/a)	The algorithm should guarantee the TNs subsequent to TN_B can meet their deadlines regardless of the mode setting of m_B. Hence, the safe mode W is not required in the algorithm
PCT	T_{cur}	Current time of the processor
Variables to be solved by the scheduler		
$s_1 \ldots s_N$	$s_1 \ldots s_B$	Start time of $TN_1 \ldots TN_B$
$m_1 \ldots m_N$	$(m_{DVFS,i}, m_{HDP,i})$, $1 \leq i \leq B$	Mode setting for $TN_1 \ldots TN_B$

In our specific case, $T_{sw,HDP} = 0$ is assumed, since HDP switching time is negligible as compared to the time granularity of scheduling. Hence, $O(m_a, m_b)$ can be simplified as follows:

$$O(m_a, m_b) = \begin{cases} 0 & \text{if } m_{DVFS,a} = m_{DVFS,b} \\ T_{sw,DVFS} & \text{if } m_{DVFS,a} \neq m_{DVFS,b} \end{cases} \quad (5.10)$$

- In the algorithm, $f_{\max}$ denotes the fastest clock frequency of all modes. Hence, it can be defined as follows:

$$f_{\max} = \frac{1}{CT[F]} \tag{5.11}$$

- The number of jobs to be scheduled (N) is equal to the size of the buffer (B), in which each job J_i represents a thread node TN_i. Not all TNs have hard deadlines, since the end of a TN can be merely an arbitrary point in the middle of the application. However, as mentioned in Sect. 5.4.4, at the start of the algorithm, an optimal schedule is drafted, in which provisional deadlines are given for all TNs without hard deadlines, serving as checkpoints of the workload process. These deadlines become additional constraints that need to be fulfilled besides the original hard deadlines. In the mathematical representation, the deadlines $DL_1, \ldots DL_B$ include both original deadlines and provisional deadlines.
- A thread node TN_i should be always ready at the end time of the previous thread node TN_{i-1}. Hence, Constraint 1 $s_i \geq R_i$ is always fulfilled. In practice, this means that all input data required for TN_i is either already stored in the processor core before TN_1 starts or provided by the computation results of previous TNs. There is no need to wait for the arrival of the data from the external components.
- In the context of dynamic scenarios, UC_i, LC_i are determined by the scenario to which TN_i belongs. In our approach, the UC and LC of each scenario are the worst and the average execution cycles of this scenario, denoted by WC_i and AC_i. These numbers are acquired from design-time process, as described in Sect. 5.4.2. By definition, the following statement is always true:

$$WC_i \geq AC_i \quad \forall 1 \leq i \leq B \tag{5.12}$$

- In the algorithm, PDL_i denotes the latest end time for TN_i to ensure that the deadlines of TN_i and the following TNs can always be met. It is the counterpart of LC_i with respect to the start time and the end time of a job. Hence, similar to LC_i, PDL_i can be defined as follows:

$$PDL_B = DL_B \tag{5.13}$$

$$PDL_i = \min\left\{PDL_{i+1} - \frac{WC_{i+1}}{f_{\max}}, DL_i\right\} \quad \forall i < B \tag{5.14}$$

Lines 10–13 of Algorithm 5.5 implement the calculation process of PDL_i. It is also worth noting that the following relation between LS_i and PDL_i can be easily proved by mathematical induction:

$$PDL_i = LS_i + \frac{WC_i}{f_{\max}} \quad 1 \leq i \leq B \tag{5.15}$$

- The TNs and mode settings before TN_1 are considered to be unknown, and the scheduler should ensure the timing guarantees regardless of the mode setting prior to TN_1. Hence, the switching time at the start of the first TN ($O(m_0, m_1)$) is defined as the maximum switching time ($T_{sw,DVFS}$) for safe scheduling.
- For the same reason as above, for the TNs subsequent to TN_B, the scheduler should guarantee that they can meet their deadlines regardless of the mode setting of TN_B. Hence, we assume that the switching at the end of TN_B occurs after D_B, and there is no need for the requirement of the safe mode W. Mathematically it is equivalent to defining $O(m_N, W) = 0$.

To avoid ambiguity, in the following sections, we will adopt the denotations of the algorithm in the proof of timing guarantees, except for $CT[m]$, $O(m_a, m_b)$, and F, which do not have equivalent denotations in the algorithm.

Based on the variable mapping as above, the hard constraints of timing guarantees in Sect. 5.5.1 can be translated in the context of the algorithm as follows:

Constraint 1: The start time of a TN cannot be earlier than its ready time. As discussed above, this constraint should be always fulfilled in our specific context.

Constraint 2: The start time of a TN cannot be earlier than the likely time of the previous TN plus switching time overhead:

$$s_{i+1} \geq s_i + AC_i \cdot CT[m_i] + O(m_i, m_{i+1}) \quad \forall i \geq 1 \tag{5.16}$$

Boundary case of the above constraint for the first job:

$$s_1 \geq T_{cur} + T_{sw,DVFS} \tag{5.17}$$

Constraint 3: A TN must end before its deadline even for the worst-case execution cycles:

$$s_i + WC_i \cdot CT[m_i] \leq DL_i \tag{5.18}$$

Constraint 4: Even when the current and future TNs pose the worst-case execution cycles, there should be enough time either (i) to switch to the fastest mode and execute all the later TNs using the fastest mode and complete them before their respective deadlines or (ii) selected modes for all jobs are such that all deadlines will be met even for the worst-case execution cycles:

$$s_i + WC_i \cdot CT[m_i] \leq \max\Big\{s_{i+1} - O(m_i, m_{i+1}), \; PDL_{i+1} - \frac{WC_{i+1}}{f_{\max}} - O(m_i, F)\Big\} \quad \forall i < B \tag{5.19}$$

5.5.3 Proof of Hard Constraints

Instead of the static schedule theorized in the model, our approach uses a dynamic schedule, which changes at run-time over the progress of the workload and is updated at the start of each TN. To ensure timing guarantees, the dynamic schedule should always fulfill the hard constraints within the buffer no matter how the schedule changes. In our proposed approach, at the start of every job, the Refined Upper Bound process of Algorithm 5.5 is invoked to formulate a schedule that strictly fulfills the hard constraints. With this schedule, we can derive the available modes for the next TN that can ensure no deadline misses for the next job and the following TNs in the buffer. Then, the next TN is executed with one of these modes until the end of the TN when the schedule is updated again. In this way, we can ensure that every mode setting for each TN is based on a schedule that fulfills the hard constraints, so the timing guarantees are ensured at any time of the workload progress.

The schedule formulated by the Refined Upper Bound process is composed of two phases, illustrated in Fig. 5.15. In the first phase, the workload runs under mode m_A, in which the speed is equal to or greater than the minimum required speed f_{mr}. Then, for the rest of the TNs, the fastest mode F is used. At the start of both phases, mode switching occurs with the worst-case switching time $T_{sw,DVFS}$. Mathematically, the schedule can be expressed as follows:

- Assume the first phase contains $TN_1 \dots TN_K$ and the second phase contains jobs $TN_{K+1} \dots TN_B$. Then, the mode setting for all jobs and the switching time overhead between jobs can be expressed as follows:

$$m_i = \begin{cases} m_A & 1 \le i \le K \\ F & i > K \end{cases} \tag{5.20}$$

$$O(m_i, m_{i+1}) = \begin{cases} T_{sw,DVFS} & \text{if } i = 0, K \\ 0 & \text{otherwise} \end{cases} \tag{5.21}$$

- In the algorithm, the interval between two DVFS switches must be at least $T_{th,DVFS}$. The boundary of the two phases is defined by this constraint. To ensure the duration of the first phase is at least $T_{th,DVFS}$, the first TN in the second phase (J_{K+1}) is defined as the first TN whose start time is at least $T_{th,DVFS} + T_{sw,DVFS}$ later than the start time of TN_1. Since from Eq. 5.28 we know that $s_{K+1} = PDL_K$, the location of TN_K is constrained by the following relations:

$$PDL_{K-1} < T_{cur} + T_{th,DVFS} + 2T_{sw,DVFS} \le PDL_K \tag{5.22}$$

The process of calculating the TN index K is implemented in Lines 15–18 of Algorithm 5.5.

- f_{mr} denotes the minimum required speed for the first phase that can still ensure the timing guarantees of all TNs. Based on Lines 19–24 of Algorithm 5.5, the minimum required speed can be expressed as follows:

$$f_{mr} = \max_{1 \leq i \leq K} \left\{ \frac{\sum_{j=1}^{i} WC_j}{DL_i - T_{cur} - T_{sw,DVFS}}, \frac{\sum_{j=1}^{K} WC_j}{PDL_K - T_{cur} - 2T_{sw,DVFS}} \right\} \tag{5.23}$$

- Based on the definition of f_{mr}, the speed of the selected mode for the first phase (m_A) must be equal to or faster than f_{mr}. This principle is implemented in Lines 11–20 of Algorithm 5.4, in which the algorithm always selects a mode faster than f_{mr} if possible. Hence, the following constraint holds:

$$CT[m_A] \leq \frac{1}{f_{mr}} \tag{5.24}$$

- In the first phase, each TN starts right after its previous TN ends, assuming each TN requires its own worst-case execution cycles. Therefore, the start time of these TNs can be expressed as follows:

$$s_1 = T_{cur} + T_{sw,DVFS} \tag{5.25}$$

$$s_i = s_{i-1} + WC_{i-1} \cdot CT[m_A] \quad 1 < i \leq K \tag{5.26}$$

From the above equations we can derive the start time of the TNs as follows:

$$s_i = T_{cur} + T_{sw,DVFS} + CT[m_A] \sum_{j=1}^{i-1} WC_j \quad 1 < i \leq K \tag{5.27}$$

- In the second phase, each TN starts at the time of the propagated deadline of the previous TN, which can be expressed as follows:

$$s_i = PDL_{i-1}, \quad K < i \leq N \tag{5.28}$$

In summary, the start time of each TN is defined in Eqs. 5.25, 5.27, and 5.28, and the mode setting of each TN is defined in Eq. 5.20. Next, we will show that the schedule always fulfills the hard constraints described in Sect. 5.5.2, and thus the timing guarantees of the algorithm are proven. The proof of Constraint 1 is dismissed since the TN is always considered ready when the previous TN ends, as mentioned in Sect. 5.5.2. The proofs of other constraints are shown as follows:

Proof of Constraint 2 The constraint can be divided into four different cases: (i) the boundary case of Eq. 5.17, (ii) $1 \leq i < K$, (iii) $i = K$, and (iv) $K < i < B$. Each of these cases will be proven, respectively, as follows:

(i) With Eq. 5.25, the constraint is automatically fulfilled:

$$s_1 = T_{cur} + T_{sw,DVFS} \geq T_{cur} + T_{sw,DVFS} \tag{5.29}$$

(ii) When $1 \leq i < K$, from Eqs. 5.20 and 5.21, we get that $m_i = m_A$ and $O(m_i, m_{i+1}) = 0$. Based on these relations and Eqs. 5.26 and 5.12, we can prove the constraint by the following derivation:

$$\begin{aligned} s_{i+1} &= s_i + WC_i \cdot CT[m_A] \\ &\geq s_i + AC_i \cdot CT[m_i] + O(m_i, m_{i+1}) \end{aligned} \tag{5.30}$$

(iii) When $i = K$, from Eqs. 5.23, 5.28, 5.24, and 5.27, we can acquire the following relations, respectively:

$$f_{mr} \geq \frac{\sum_{j=1}^{K} WC_j}{PDL_K - T_{cur} - 2T_{sw,DVFS}} \tag{5.31}$$

$$PDL_K = s_{K+1} \tag{5.32}$$

$$\frac{1}{CT[m_A]} \geq f_{mr} \tag{5.33}$$

$$\sum_{i=1}^{K} WC_i = \sum_{i=1}^{K-1} WC_i + WC_K = \frac{s_K - T_{cur} - T_{sw,DVFS}}{CT[m_A]} + WC_K \tag{5.34}$$

Then from the above relations we can derive

$$\frac{1}{CT[m_A]} \geq f_{mr} \geq \frac{s_K - T_{cur} - T_{sw,DVFS} + WC_K \cdot CT[m_A]}{CT[m_A](s_{K+1} - T_{cur} - 2T_{sw,DVFS})} \tag{5.35}$$

$$s_{K+1} - T_{cur} - 2T_{sw,DVFS} \geq s_K - T_{cur} - T_{sw,DVFS} + WC_K \cdot CT[m_A] \tag{5.36}$$

$$s_{K+1} \geq s_K + WC_K \cdot CT[m_K] + T_{sw,DVFS} \tag{5.37}$$

$$s_{K+1} \geq s_K + AC_K \cdot CT[m_K] + O(m_K, m_{K+1}) \tag{5.38}$$

(iv) When $K < i < N$, from Eqs. 5.20 and 5.21, we get that $m_i = F$ and $O(m_i, m_{i+1}) = 0$. Based on this relation and Eqs. 5.14, 5.28, and 5.12, we can prove the constraint by the following derivation:

$$PDL_{i-1} \leq PDL_i - WC_i \cdot CT[F] \tag{5.39}$$

$$s_i \leq s_{i+1} - WC_i \cdot CT[F] \tag{5.40}$$

$$s_{i+1} \geq s_i + WC_i \cdot CT[F] \tag{5.41}$$

$$s_{i+1} \geq s_i + AC_i \cdot CT[m_i] \tag{5.42}$$

Proof of Constraint 3 The constraint can be divided into four different cases: (i) $1 \leq i < K$, (ii) $i = K$, (iii) $K < i < B$, and (iv) the boundary case of Eq. 5.5. Each of these cases will be proven, respectively, as follows:

(i) When $1 \leq i < K$, from Eqs. 5.23, 5.24, and 5.27, we can acquire the following relations, respectively:

$$f_{mr} \geq \frac{\sum_{j=1}^{i} WC_j}{DL_i - T_{cur} - T_{sw,DVFS}} \tag{5.43}$$

$$\frac{1}{CT[m_A]} \geq f_{mr} \tag{5.44}$$

$$\sum_{j=1}^{i} WC_j = \frac{s_{i+1} - T_{cur} - T_{sw,DVFS}}{CT[m_A]} \tag{5.45}$$

From the above relations we can derive

$$f_{mr} \geq \frac{s_{i+1} - T_{cur} - T_{sw,DVFS}}{CT[m_A](DL_i - T_{cur} - T_{sw,DVFS})} \tag{5.46}$$

$$f_{mr} \geq \frac{f_{mr}(s_{i+1} - T_{cur} - T_{sw,DVFS})}{DL_i - T_{cur} - T_{sw,DVFS}} \tag{5.47}$$

$$DL_i - T_{cur} - T_{sw,DVFS} \geq s_{i+1} - T_{cur} - T_{sw,DVFS} \tag{5.48}$$

$$DL_i \geq s_{i+1} = s_i + WC_i \cdot CT[m_i] \tag{5.49}$$

(ii) When $i = K$, from Eqs. 5.14 and 5.28 we can derive

$$DL_K \geq PDL_K = s_{K+1} \tag{5.50}$$

Then, from Eq. 5.37 we can further derive

$$DL_K \geq s_K + WC_K \cdot CT[m_K] + T_{sw,DVFS} \tag{5.51}$$

$$DL_K \geq s_K + WC_K \cdot CT[m_K] \tag{5.52}$$

(iii) When $K < i < N$, from Eq. 5.14 we can derive

$$PDL_i - \frac{WC_i}{f_{\max}} \geq PDL_{i-1} \tag{5.53}$$

$$PDL_i \geq PDL_{i-1} + \frac{WC_i}{f_{\max}} \tag{5.54}$$

$$DL_i \geq PDL_{i-1} + \frac{WC_i}{f_{\max}} \tag{5.55}$$

From Eqs. 5.28 and 5.20, we know that $PDL_{i-1} = s_i$ and $m_i = F$, so we get

$$DL_i \geq s_i + WC_i \cdot CT[m_i] \tag{5.56}$$

(iv) When $i = B$, from Eqs. 5.13 and 5.14 we can derive

$$PDL_B - \frac{WC_B}{f_{\max}} \geq PDL_{B-1} \tag{5.57}$$

$$PDL_B \geq PDL_{B-1} + \frac{WC_B}{f_{\max}} \tag{5.58}$$

$$DL_B \geq PDL_{B-1} + \frac{WC_B}{f_{\max}} \tag{5.59}$$

From Eqs. 5.28 and 5.20, we know that $PDL_{B-1} = s_B$ and $m_B = F$, and in Sect. 5.5.2 we define $O(m_B, W) = 0$. Hence, we can acquire

$$DL_B \geq s_B + WC_B \cdot CT[m_B] + O(m_B, W) \tag{5.60}$$

Proof of Constraint 4 In the process of the proof of Constraint 2, from Eqs. 5.30, 5.37, and 5.41, we can observe that the following statement is true for all $i < B$:

$$s_{i+1} \geq s_i + WC_i \cdot CT[m_i] + O(m_i, m_{i+1}) \tag{5.61}$$

Hence, we can derive

$$\begin{aligned} s_i + WC_i \cdot CT[m_i] &\leq s_{i+1} - O(m_i, m_{i+1}) \\ &\leq \max\left\{s_{i+1} - O(m_i, m_{i+1}), PDL_{i+1} - \frac{WC_{i+1}}{f_{\max}} - O(m_i, F)\right\} \end{aligned} \tag{5.62}$$

Since we have proven that all hard constraints are fulfilled in the dynamic schedule kept in the algorithm, we prove that the approach can ensure the timing guarantees.

5.6 Applicability to Other Architectures, Technologies, and Hardware

Although the presented approach is implemented on a single-core processor, the principles of multi-timescale knobs and dynamic scenarios are generic. Therefore, they can be instantiated to develop mitigation approaches of other architectures. Each architecture has its unique knobs, such as core mapping in multicore processors and thread mapping in GPUs. Nevertheless, in Sects. 5.2 and 5.3 we have shown the method of exploring time-energy trade-offs and switching overheads of the knobs. The principles of the method can be used on the other knobs to know their capability and limitation to be exploited in mitigation approaches. Once the knobs are explored, the DS principles described in Sects. 5.4.1 and 5.4.3 can be instantiated to develop a mitigation approach that can tackle the architecture-specific variability sources, such as cache protocol in multicore processors and thread management in GPUs.

Moreover, the principles of the presented approach are not specific to any technology, so they can be applied to different technologies. To transfer the approach to a new technology, one should explore the variety of the devices and circuits of the technology which can result in performance-energy trade-offs, such as voltage scaling in CMOS technology. The trade-offs can be utilized to design knobs. With a variety of knobs of different ranges and switching time, the mitigation approach can be developed on the same principles.

Still, also other reliability requirements and quality-of-service constraints can be applicable, as already mentioned. We believe that the main principles that are proposed here, namely the use of additional fast knobs like HDP, and especially the use of guarantee-oriented dynamic scenarios, can be reused to meet also the other QoS requirements, on top of timing. In that case, the objective function will become a guarantee in an N-dimensional Pareto space. Some hints of how to combine performance mitigation with other objectives are provided in Chap. 7.

In summary, the instantiation of the multiple-knob and DS principles to other architectures and technologies is a whole new research direction that is yet to be explored. Hence, research should be performed in the future to close the gap of timing guarantees for modern complicated architectures and advanced technologies.

5.7 Summary

The workload-dependent multi-timescale mitigation approach is developed for ensuring timing guarantees of highly dynamic workloads. The proposed approach follows the basic structure of a run-time mitigation approach which consists of monitors, run-time controller, and system knobs. However, the approach is improved by two salient features, so that the full timing guarantees can be realized. First, a new knob, HDP, is introduced, which enables multiple datapaths of different speeds in the processor core. HDP can be used to tackle fast variability of the nanosecond level. By incorporating both the HDP and DVFS, the timescale of the mitigation can be pushed down to the sub-millisecond level, fulfilling the real-time requirements necessary for ensuring timing guarantees. Second, the controlling algorithm is instantiated from the dynamic scenario (DS) concept, which allows the upper bounds of execution time to be derived at run-time using scenarios created at design-time profiling. The DS-based algorithm ensures the timing requirements are always met, which is proven by mathematical deduction. The approach we proposed here is the first mitigation approach for performance variability which can ensure timing guarantees to the best of our knowledge.

References

1. Balkind, J., Chang, T.J., Jackson, P.J., Tziantzioulis, G., Li, A., Gao, F., Lavrov, A., Chirkov, G., Tu, J., Shahrad, M., Wentzlaff, D.: OpenPiton at 5: a nexus for open and agile hardware design. IEEE Micro **40**(4), 22–31 (2020). https://doi.org/10.1109/MM.2020.2997706
2. Bambagini, M., Marinoni, M., Aydin, H., Buttazzo, G.: Energy-aware scheduling for real-time systems: a survey. ACM Trans. Embed. Comput. Syst. **15**(1) (2016). https://doi.org/10.1145/2808231
3. Catthoor, F., Basten, T., Zompakis, N., Geilen, M., Kjeldsberg, P.G.: System-Scenario-based Design Principles and Applications. Springer International Publishing, Dordrecht (2020)
4. Gheorghita, S.V., Palkovic, M., Hamers, J., Vandecappelle, A., Mamagkakis, S., Basten, T., Eeckhout, L., Corporaal, H., Catthoor, F., Vandeputte, F., Bosschere, K.D.: System-scenario-based design of dynamic embedded systems. ACM Trans. Des. Autom. Electron. Syst. **14**(1) (2009). https://doi.org/10.1145/1455229.1455232
5. Gwennap, L.: MIPS R10000 uses decoupled architecture. Microprocess. Rep. **8**(14), 18–22 (1994). https://doi.org/10.1109/TVLSI.2022.3177861
6. Hammari, E., Catthoor, F., Kjeldsberg, P.G., Huisken, J., Tsakalis, K., Iasemidis, L.: Identifying data-dependent system scenarios in a dynamic embedded system. In: The 2012 International Conference on Engineering of Reconfigurable Systems and Algorithms (ERSA'12), pp. 70–76 (2012)
7. Hu, Z., Buyuktosunoglu, A., Srinivasan, V., Zyuban, V., Jacobson, H., Bose, P.: Microarchitectural techniques for power gating of execution units. In: Proceedings of the 2004 International Symposium on Low Power Electronics and Design (IEEE Cat. No.04TH8758), pp. 32–37 (2004). https://doi.org/10.1145/1013235.1013249
8. Ma, Z., Marchal, P., Scarpazza, D.P., Yang, P., Wong, C., Gómez, J.I., Himpe, S., Ykman-Couvreur, C., Catthoor, F. (eds.): Systematic Methodology for Real-Time Cost-Effective Mapping of Dynamic Concurrent Task-Based Systems on Heterogenous Platforms. Springer Netherlands, Dordrecht (2007)
9. Munaga, S., Catthoor, F.: Systematic design principles for cost-effective hard constraint management in dynamic nonlinear systems. Int. J. Adapt. Resilient and Autonomic Syst. **2**(1), 18–45 (2011). https://doi.org/10.4018/jaras.2011010102
10. Sato, T., Arita, I.: Execution latency reduction via variable latency pipeline and instruction reuse. In: Sakellariou, R., Gurd, J., Freeman, L., Keane, J. (eds.) Euro-Par 2001 Parallel Processing, pp. 428–438. Springer Berlin Heidelberg, Berlin (2001)
11. Scordino, C., Lipari, G.: A resource reservation algorithm for power-aware scheduling of periodic and aperiodic real-time tasks. IEEE Trans. Comput. **55**(12), 1509–1522 (2006). https://doi.org/10.1109/TC.2006.190

12. Waterman, A., Asanović, K., Hauser, J. (eds.): The RISC-V Instruction Set Manual, Volume II: Privileged Architecture Document Version 20211203. RISC-V International, New York (2021)
13. Yakimets, D., Bardon, M.G., Jang, D., Schuddinck, P., Sherazi, Y., Weckx, P., Miyaguchi, K., Parvais, B., Raghavan, P., Spessot, A., Verkest, D., Mocuta, A.: Power aware FinFET and lateral nanosheet FET targeting for 3 nm CMOS technology. In: 2017 IEEE International Electron Devices Meeting (IEDM), pp. 20.4.1–20.4.4 (2017). https://doi.org/10.1109/IEDM.2017.8268429
14. Zaruba, F., Benini, L.: The cost of application-class processing: energy and performance analysis of a Linux-ready 1.7-GHz 64-bit RISC-V core in 22-nm FDSOI technology. IEEE Trans. Very Large Scale Integr. VLSI Syst. **27**(11), 2629–2640 (2019). https://doi.org/10.1109/TVLSI.2019.2926114

Chapter 6
Workload-Dependent Multi-Timescale Mitigation Approach for Performance Variability: Experiments

Abstract In this chapter, we experiment on the effectiveness of the workload-dependent multi-timescale mitigation approach for performance variability, with a cycle-accurate simulation flow we have developed. The targeted applications for the approach are highly dynamic workloads with strict real-time requirements. The state-of-the-art approaches fail to provide practically usable full guarantees for these workloads, so we consider them the most suited benchmarks. We select two representative workloads from widely different application domains. One is the lower sub-band quantization block of the ADPCM encoding application from TACLeBench Falk et al (TACLeBench: a benchmark collection to support worst-case execution time research. In: Schoeberl M (ed) 16th International Workshop on Worst-Case Execution Time Analysis (WCET 2016). OpenAccess Series in Informatics (OASIcs), vol 55, pp 2:1–2:10. Schloss Dagstuhl–Leibniz-Zentrum für Informatik, Dagstuhl, 2016). The other one is the JPEG decoding application from MiBench Guthaus et al (MiBench: a free, commercially representative embedded benchmark suite. In: Proceedings of the Fourth Annual IEEE International Workshop on Workload Characterization, pp 3–14, 2001. https://doi.org/10.1109/WWC.2001.990739). Both workloads are typical real-time signal and data processing applications, where the input of workload is a sequence of frames. Deadlines for processing consecutive frames are periodic, where the interval between two consecutive deadlines is called *frame period*.

In conventional real-time scheduling, the finest unit is the job of processing a single frame, but in our approach, more fine-grained mitigation can be achieved by cutting the job of processing a frame into multiple TNs, each of which has an execution time of tens to hundreds of microseconds. The execution time of these workloads is highly dynamic with different input data. Therefore, they are representative of the highly dynamic workloads that our approach targets.

This chapter is structured as follows: Sect. 6.1 elaborates on the experiment setup, in which the cycle-accurate simulation flow is introduced. Then, Sects. 6.2 and 6.3 discuss the case studies with the workloads of lower-band quantization of ADPCM encoding and JPEG decoding. The comparison of the results of the case studies is discussed in Sect. 6.4.

J.-Y. Lin et al., *Workload Dependent Mitigation Approaches for Performance Variability*, https://doi.org/10.1007/978-3-031-85624-2_6

6.1 Experimental Setup

We develop a simulation flow that can simulate the entire process of mitigation approaches, including both the design-time and run-time processes. In the simulation flow, cycle-accurate processor simulation of execution time and energy consumption is used. The details of the simulation flow are elaborated in the following sections.

6.1.1 Processor Simulation Flow

Accurate simulation of the execution time and energy consumption of the processor is required in many steps of the simulation of the DS-based methodology, such as the design-time profiling and the run-time workload benchmarking. Therefore we established a cycle-accurate processor simulation flow for this purpose, which is shown in Fig. 6.1.

In the processor simulation flow, the energy consumptions of different HDP modes are simulated separately. To accurately simulate the energy consumption of HDP, first we create the physical design of each ALU of the HDP. The circuit design and physical design of the HDP knob are already detailed in Sect. 5.2. Then, each ALU of the HDP has gone through the entire digital design flow, including logic synthesis and place and route, to generate the full physical design. The netlists

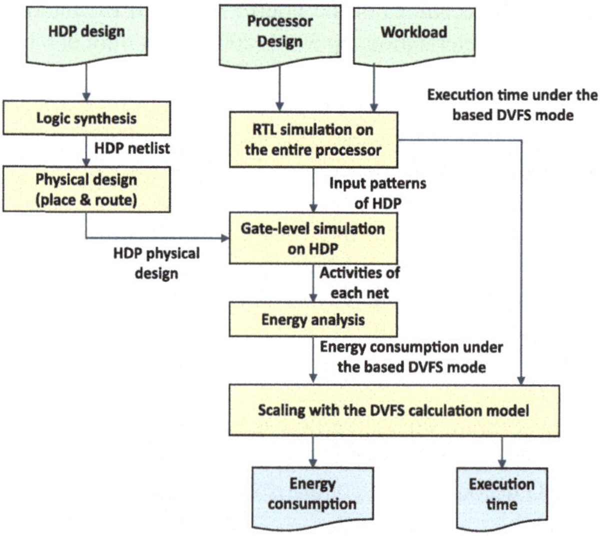

Fig. 6.1 The processor simulation flow of benchmarking the execution time and HDP energy consumption for running a certain workload

of the physical designs are further used in gate-level simulation to obtain detailed activities inside the HDP when running a workload.

Then, we perform RTL-level simulation on the RTL design of the entire processor core, running the target workload. Execution time results are acquired in this step in cycle accuracy. Meanwhile, the input vectors of the HDP are extracted during the simulation. Then, gate-level simulation is performed on the netlist of each ALU of the HDP. The input vectors from the RTL simulation are applied to the ALU design in the simulation to acquire the activities of each net in the ALUs. The activities record every rising and falling transition of each net during the workload at the picosecond level. Afterward, these activities are used by the commercial power analysis tool to compute the energy consumption of the ALUs.

In our energy simulation, the time and dynamic energy overheads of HDP switching are not incorporated, since the HDP switching time is only tens of cycles, which is only about 0.1% of execution time in our sub-millisecond mitigation context, so it would only have negligible influence on time and energy. Moreover, the energy overheads from the idle ALUs are also very small compared to the energy of the active ALU. Even though the duration of power-gating can be as short as tens of microseconds, it is still thousands of times longer than the break-even point, which is approximately ten cycles [3]. Therefore, the energy consumption of the idle ALUs is also left out of the analysis.

On the other hand, the timing and energy consumption of different DVFS modes are calculated from the ones of the base DVFS mode with a timing and energy calculation model. We create five DVFS modes, with voltages ranging from 0.5 V to 0.9 V. The clock frequency of each mode is calculated by static timing analysis on the ALU design, while the energy consumption of each mode is calculated by scaling from the energy consumption of the 0.7 V mode, which is the base mode of our DVFS knob. The energy scaling is based on the V^2 scaling relation according to Eq. 3.2. Here, we do not consider static energy, since in simulation we found it smaller than 3% of the dynamic energy due to all the leakage reduction techniques (high-V_{th} library, power gating). A summary of the DVFS model is shown in Table 6.1, and the speed-energy relations of all available DVFS and HDP modes are shown in Fig. 5.8. In addition, we can realistically assume that the switching time of

Table 6.1 Summary of DVFS model

	Frequency (GHz)			Energy[a] (nJ)		
Voltage (V)	Fast ALU	Mid ALU	Slow ALU	Fast ALU	Mid ALU	Slow ALU
0.9	4.67	4.46	3.33	22.37	19.01	16.07
0.8	4.24	3.98	2.97	17.68	15.02	12.69
0.7 (base mode)	3.69	3.39	2.46	13.53	11.50	9.72
0.6	2.80	2.08	1.60	9.94	8.45	7.14
0.5	1.79	0.98	0.79	6.90	5.87	4.96

[a] The actual energy consumption is workload-dependent by the cycle-accurate simulation flow. The numbers shown here are examples of simulated with a JPEG encoding workload.

DVFS is a constant of 100 μs regardless of the modes the processor switches from and to. This switching time is representative of the empirical numbers measured from modern processors [4].

6.1.2 Overall Simulation Flow

Figure 6.2 shows the overall simulation flow for the experiments of the mitigation approaches. Both the design-time scenario clustering and the run-time scheduling parts in the approach are simulated. During the simulation of design-time scenario clustering, we profile the workload with different input files to acquire the statistics of execution time and energy, using the processor simulation flow described in the last section. Then, based on the execution flow of the workload, the profiling traces are segmented into TNs of tens to hundreds of microseconds in length. Once the execution time and energy of all TNs are acquired, we analyze the correlation between these data and application variables to search for suitable monitoring parameters for scenario clustering. Scenarios are generated based on the monitoring parameters, and the statistics of each scenario are calculated and stored in a look-up table. The execution flow analysis, TN segmentation, discovery of monitoring parameters, and scenario clustering are all performed manually.

During the simulation of run-time scheduling, arbitrary inputs are selected for testing the algorithm. Then, we benchmark the workload with the inputs to acquire

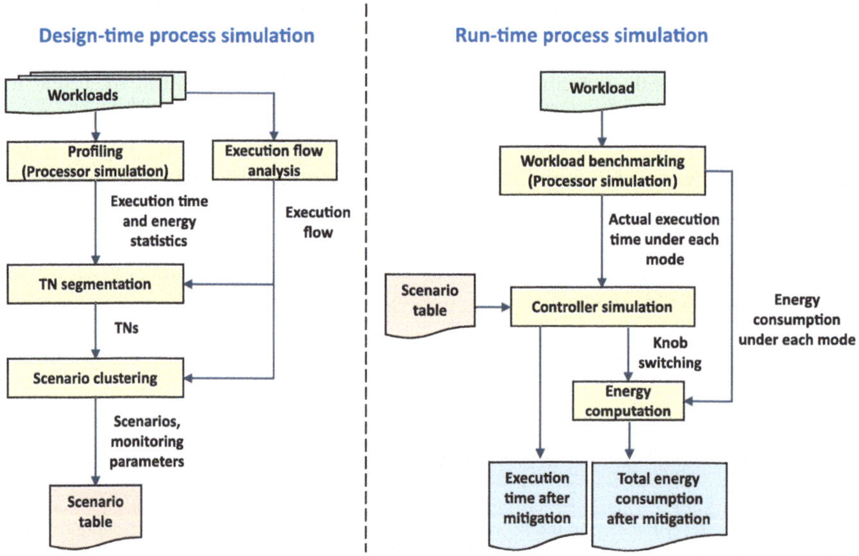

Fig. 6.2 The cycle-accurate simulation flow to acquire the execution time and energy consumption when running a workload under mitigation

their execution time and energy under different modes of the knobs, using the processor simulation flow described in the last section. By applying the inputs of different frame contents, performance variability is injected, leading to differences in execution time for processing each frame. The exact execution time of each TN is captured by the processor simulation. These results are read by a controller simulator in which the adaptive control algorithm is implemented in the Perl language. With the scenario table and the benchmarking results, the simulator derives the knob-switching events issued by the controller along the entire workload. The results of these knob-switching events are combined with the execution time results and the energy analysis results to calculate the execution time and the total energy consumption after mitigation.

In the experiment, we compare six run-time mitigation approaches of different algorithms and knobs. In terms of the algorithms, we select three different configurations, each of which is representative of an algorithm type: best-effort (BE), WCET-based, and DS-based. The BE algorithms are represented by the algorithm of Pering et al. [5], in which the prediction of execution cycles is the exponential moving average of the executed cycles of the past frames. Though this algorithm cannot ensure timing guarantees, it represents the algorithm of the best energy consumption one can achieve. WCET-based algorithms are represented by the cycle-conserving algorithm of Pillai et al [6]. This configuration represents the conventional algorithm for timing guarantees.

In terms of the knobs, we have two different configurations. The first one is the single-DVFS configuration, which only has the DVFS knob. The only ALU in the processor is assumed to be the same as the fast ALU in the HDP. This configuration represents the conventional approach which only has the DVFS knob and one datapath designed for high speed. The BE and the WCET-based algorithms of this configuration switch DVFS modes only at the beginning of each frame, regardless of the TNs. On the other hand, the DS-based algorithm can switch at the start of each TN, while still following the restriction that the interval between two DVFS switches must be longer than $T_{\mathrm{th,DVFS}}$. The second one is the multi-timescale configuration, which uses both the DVFS and the HDP knob. This configuration represents the proposed approach. For BE and the WCET-based algorithms, their multi-timescale versions are implemented according to the flow in Fig. 5.14. Therefore, they can switch the knobs at the start of each TN. Still, their execution time predictions are frame-based and are unaware of the run-time situation, so they cannot fully take advantage of the finer timescale. Combining different configurations of the knobs and the algorithms, we have six different approaches for comparison. By comparing the results of different configurations, we can decouple the impacts from two different features of our methodology: either from the knobs or from the algorithm.

6.2 Case Study I: Lower-Band Quantization of ADPCM Encoding Application

We used the lower sub-band quantization block of the ADPCM-encoding application from TACLeBench [1] as our first workload. Adaptive Differential Pulse Code Modulation (ADPCM) is an audio wave encoding method which is widely used in speech codec, featuring sound compression with variable quantization steps. Figure 6.3 shows the block diagram of the encoding progress of the G.722 standard of ADPCM [7]. In a nutshell, the algorithm takes frames of audio signal samples as the inputs and generates frames of compressed bit streams of the audio signals as the outputs. In the encoding process, the audio signals are first separated into features of high-frequency and low-frequency sub-band signals. For both sub-bands, an adaptive predictor predicts the incoming samples based on the previous samples. The real incoming samples are compared with the predicted samples, and the error of prediction is recorded and then compressed by an adaptive quantizer. The quantized values of prediction errors of both sub-bands are then aggregated to be the output stream.

Here, we assume a workload of the lower sub-band quantization block of ADPCM encoding. Though the original application includes the entire ADPCM encoding flow, we only experiment on the quantization block of the low sub-band, which is the most dynamic part of the flow, to maximize the dynamism of the workload. It is feasible in practice to run the quantization block alone since one can easily modify the ADPCM encoding application to parallelize the quantization part from the other parts. Technically, we simulate the entire ADPCM encoding application but only record the execution time and energy consumption of the lower sub-band quantization block. The execution time and energy consumption of different frames are summed up to represent the case that the quantization parts are run alone. The inputs of the application are audio wave samples of 22050 Hz sample rate, as illustrated in Fig. 6.4. The frame size of the audio stream is 2000 samples.

6.2.1 *Design-Time Scenario Clustering*

First, we profile the lower-band quantization workload with the inputs of six different audio files of 10-second long to obtain the statistics of the execution time and energy consumption of the workload. These audio files are segments of music songs of different genres. Then, we divide the profile traces in a way that each TN represents the processing of 100 samples, which means that processing one frame requires 20 TNs. In this way, the execution time of each TN is around tens to hundreds of microseconds. Figure 6.5 shows the distribution of the number of instructions of all profiled TNs and the WCET condition from program flow analysis. From the distribution, we can observe that the workload is highly dynamic.

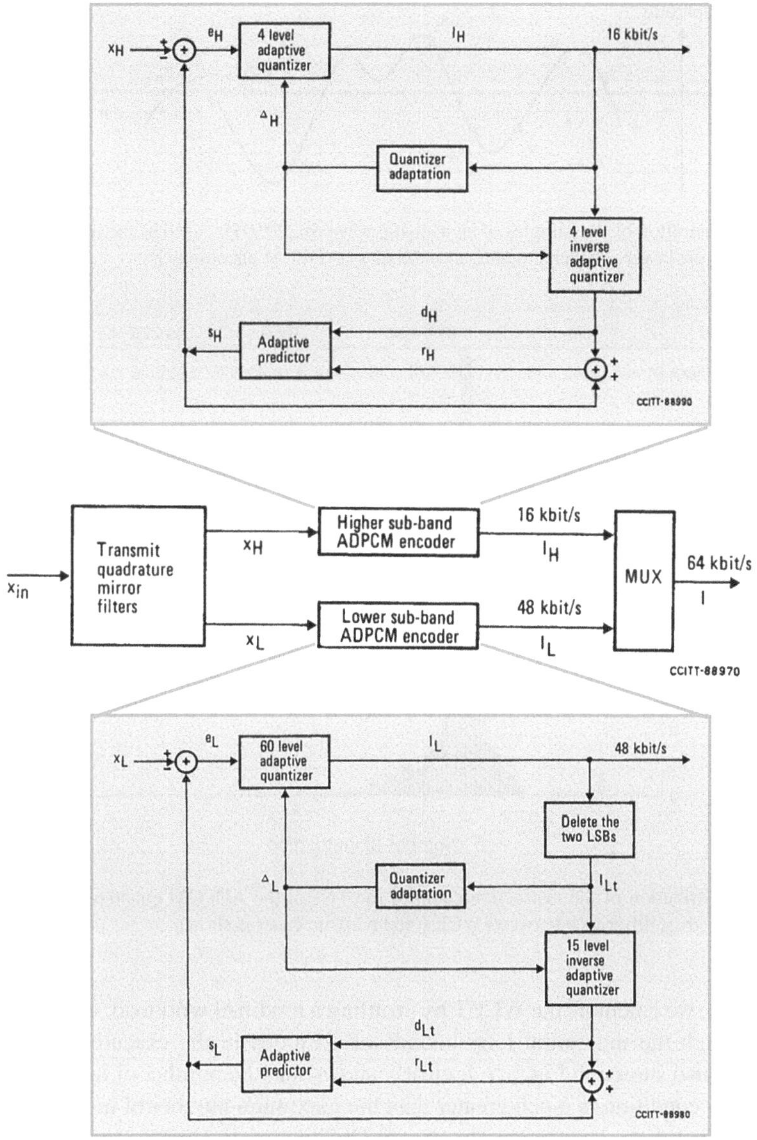

Fig. 6.3 Block diagram of the encoding process of ADPCM G.722 standard [7]

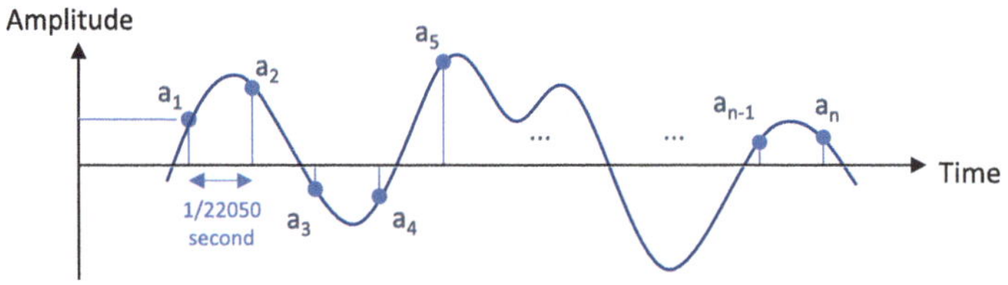

Fig. 6.4 Illustration of the samples of an acoustic wave of 22050 Hz, used as the input data of the workload of the lower sub-band quantization block of ADPCM encoding

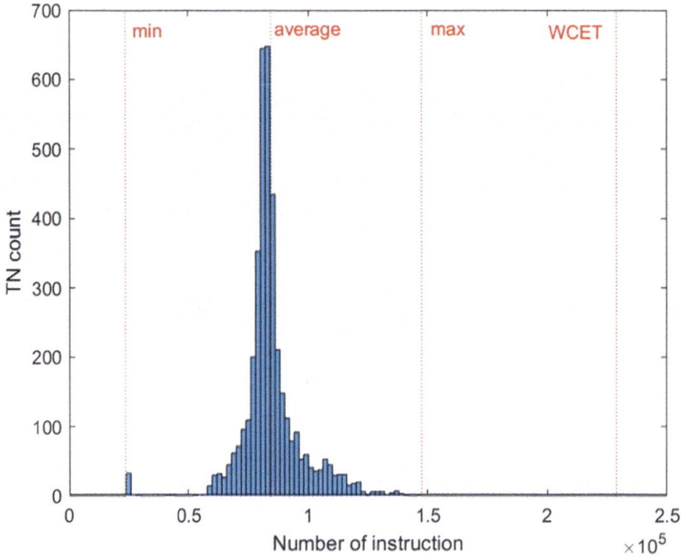

Fig. 6.5 Distribution of the instruction count of the TNs of the ADPCM quantization workload, in which the great difference between WCET and realistic cases is shown

In addition, we calculate the WCET by profiling a modified workload, which always goes through the maximum loop bounds of all loops in the execution flow. The WCET is also shown in Fig. 6.5. It clearly shows that the number of instructions of the WCET condition is much greater than the maximum number of instructions we observe. Therefore, it demonstrates that WCET is too pessimistic to represent the real execution time.

In scenario clustering, we use the *volume* as the monitoring parameter. Assuming the amplitudes of all acoustic wave samples processed in a TN are denoted by $a_1, a_2, ..., a_n$, then the volume of this TN is defined as the equation below:

$$Volumn = \sum_{i=1}^{n} |a_i| \tag{6.1}$$

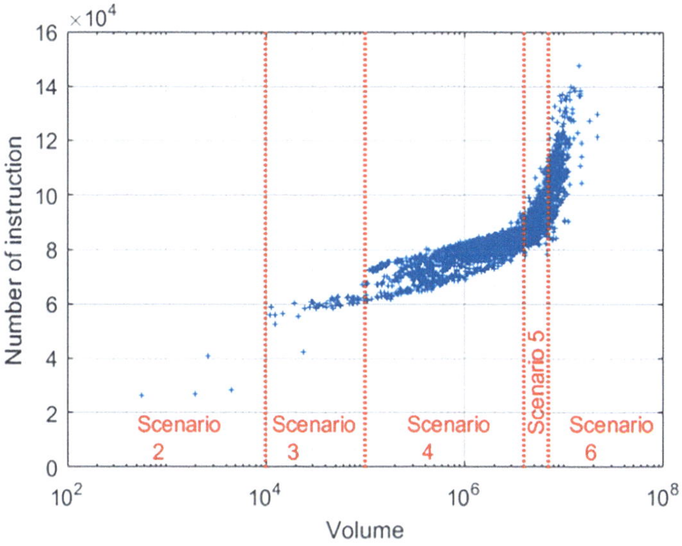

Fig. 6.6 The relation between the volume and the number of instructions for each TN. TNs are clustered into scenarios by volume

Table 6.2 Scenario look-up table for the ADPCM quantization application

Scenario	Volume range	Avg. cycle count	Max. cycle count
1	0	42276	42276
2	$1 - 10^4$	56180	76640
3	$10^4 - 10^5$	100451	124669
4	$10^5 - 4\times10^6$	147584	169068
5	$4\times10^6 - 7\times10^6$	164788	206888
6	$>7\times10^6$	202253	265514

In the profiling results, the volume is shown to have a high correlation to the number of instructions, as shown in Fig. 6.6, which makes the volume suitable to be the monitoring parameter. Then, six scenarios are created manually based on the understanding of the workload. Each scenario corresponds to a different range of the volume. Scenario 1 corresponds to the TNs of zero volume, a special case in which all samples are zero. The zero-volume TNs are frequent in the quiet periods of the audio file, which justifies the need for a specific scenario for these TNs. Scenarios 2 and 6 are designed to include the TNs of extremely low and high volumes, respectively. Scenarios 3–5 include the TNs of normal volume ranges. They are separated in a way that the difference in the number of instructions in each scenario is limited by approximately 3×10^{-4}. After the scenarios are created, the look-up table of the statistics of these scenarios is generated, shown in Table 6.2.

6.2.2 Run-Time Mitigation Results

At run-time, we selected an arbitrary audio wave of a music song with 500 samples and repeated it ten times to form the input of the workload. By applying the input, the performance variability is injected, in which the execution flow of processing each frame depends on the content of the samples. As shown in Fig. 6.5, the different input contents lead to variability in execution time, and the exact execution time cannot be calculated before execution. It requires 1000 TNs to run the whole workload, and the total execution time is 34 ms under the 0.7 V DVFS mode and the fast HDP mode. For each TN encountered at run-time, the controller calculates the volume of the TN and then references the look-up table (Table 6.2) to find the scenario of the TN. As a result, the controller obtains the average cycle count and the upper bound cycle count of the TN from the table. These cycle count data are used by the scheduling algorithm (Algorithm 5.4) to derive the suitable modes to run each TN in the workload.

Figure 6.7 shows the selected DVFS and HDP modes during the entire workload of the 1.0-ms frame period case, and Fig. 6.8 shows the traces of slack of all mitigation approaches. The slack is calculated as the amount of time by which a TN ends before its deadline. The optimal behavior of the slack trace is always to be positive and as close to zero as possible, which means that the processor runs at the slowest possible speed without any deadline violations. For the BE approaches, the slack is close to zero all along the workload, which is near-optimal for energy saving. However, as shown in Fig. 6.8, their slacks can sometimes be negative, which means potential deadline violations. Therefore, these methods cannot ensure timing guarantees.

For the approaches that can ensure timing guarantee (WCET and DS), we identify three phases of mitigation over the entire workload from the traces of slack. First, at the beginning of the workload, fast modes are used to accumulate slack to ensure a safe margin to avoid deadline violations. Then, when sufficient slack is secured to ensure timing guarantees, the processor runs at a speed at which the slack keeps at a stable level. In this phase, the processor switches among neighboring modes to fine-tune the running speed or react to small variability. Finally, when the workload is about to finish, the processor exploits the slack to save energy by running in slow modes.

While all four curves generally follow the three phases of mitigation, their behaviors are quite different. Comparing the traces of different knob configurations, we observe that the traces of the single-DVFS configurations could have sharp turns, such as the turn at about 4.6×10^{-7} cycles of the single-DVFS DS curve. These turns indicate occurrences of DVFS switching which causes dramatic changes in processor speed. As a result, the trend of slack increasing or decreasing also changes dramatically before and after the DVFS switching. It is worth noting that depending on the decisions of the algorithm, there might not be any DVFS switching during the whole workload. In this case, there would not be any sharp turns in the curve, such as the single-DVFS BE configuration in the result. However, these curves are

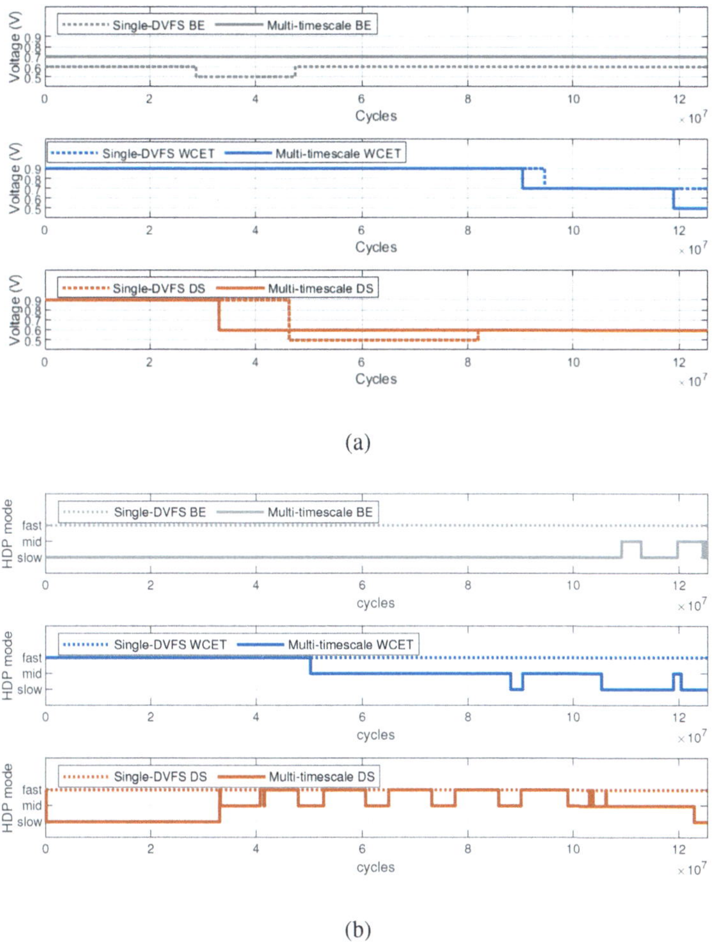

Fig. 6.7 Selected modes over the entire run of the ADPCM quantization workload with the 1.0-ms frame period. (**a**) DVFS mode and (**b**) HDP mode

still far from the ideal behavior that the slack should be kept at a level that is only required for timing guarantees. These characteristics show that with the coarse-grained modes and long intervals between switchings of DVFS, the control of the slack is far from optimal.

In contrast, the traces of the multi-timescale configurations are much smoother and kept at a constant slack level most of the time. This shows that the fine-grained modes and the small timescale enabled by HDP enable the controller to precisely control the slack and immediately tackle the variability. Therefore, the workload progress can be kept at a stable slack required for timing guarantees.

Furthermore, comparing the traces of the two approaches of the DS-based algorithm shows that the multi-timescale one requires less slack to ensure timing

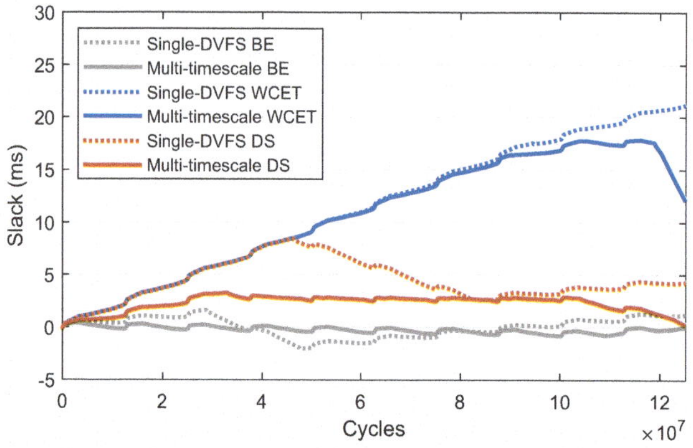

Fig. 6.8 Trace of slack over the entire run of the ADPCM quantization workload with the 1.0-ms frame period

guarantees, which results in shorter time and slower modes in the first phase. Also, the multi-timescale one starts the third phase earlier, so it can exploit more slack by the end of the workload. These behaviors show the benefits of multi-timescale mitigation in providing more energy reduction.

Moreover, the benefit of the DS-based algorithm is shown in Figs. 6.7a and 6.8 by comparing the results of the approaches with the single-DVFS configuration. The approach with the DS-based algorithm starts the second phase much earlier than the one with the WCET-based algorithm. This shows that the DS-based algorithm requires less slack to ensure timing guarantees due to the tight upper bound of execution time. The smaller slack indicates that the processor runs at a slower speed, which is beneficial to energy saving. Finally, combining the benefits of both features, the proposed approach has a smooth mitigation process over the whole workload with only the least amount of slack required to be secured. Its curve is always positive and close to zero, which is near the optimal behaviors we described above. This means it is the best approach among all in keeping the processor at the slowest possible speed without violating any deadlines.

Figure 6.9 shows the energy consumption of different approaches under different time-criticality situations, represented by different frame periods. The energy differences among all approaches are more prominent in the middle of the curves. At the left end of the curves, the timing is so strict that all approaches use the fastest mode most of the time, while at the right end of the curves, the timing is so relaxed that all methods use the slowest mode most of the time. In these conditions, the choice of the mitigation approach is less relevant.

Comparing different methods, Fig. 6.9 shows that the proposed approach has an energy saving of 32% on average and 46% on maximum, compared to the conventional WCET-based single-DVFS approach. Also, the benefit is significant in all time-criticality situations. As compared to the single-DVFS BE approach, which is near-optimal for energy saving, the energy consumption of the proposed

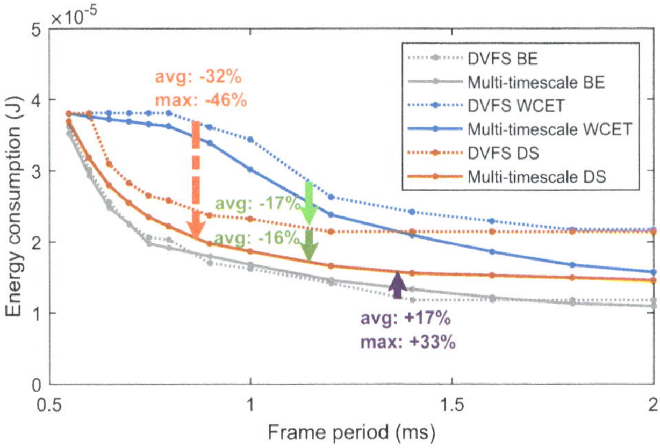

Fig. 6.9 ADPCM quantization workload energy consumption of different methods under different time-criticality conditions

approach is larger by only 17% on average and 33% on maximum. Regarding the timing guarantees, the single-DVFS BE method misses 13% deadlines on average, while the proposed approach has zero deadline misses. Since the BE approaches cause potential deadline misses, it is not an available solution for time-critical and time-sensitive applications. For these applications, the proposed approach is the best solution, in which a small energy overhead is a reasonable price to pay to ensure the timing guarantees.

The separate contributions of each feature can also be derived from Fig. 6.9. The benefit of the DS-based algorithm can be evaluated from the comparison between the single-DVFS WCET-based approach and the single-DVFS DS-based approach. It shows that the algorithm improves energy consumption by 17% on average. Its benefit is more prominent in the time-critical region, since its advantage of deriving a tighter execution time upper bound is more important in the time-critical situation. The benefit of the multi-timescale approaches using multiple knobs can be evaluated from the comparison between the single-DVFS DS-based method and the multi-timescale DS-based method. It shows the improved knob configuration saves extra 16% energy on average. Its benefit is more prominent when the timing is loose, since its advantage of allowing the processor to run in slower modes for a longer time is more prominent in the non-time-critical situation. In conclusion, it is the combined effect of both features which makes our approach have a significant energy saving in all time-criticality situations.

6.3 Case Study II: JPEG Decoding Application

We used the JPEG-decoding application from MiBench [2] as the second workload. JPEG is a widely used image compression standard. A JPEG encoder can compress a bitmap digital image into a shorter bit stream, while a JPEG decoder can transform the bit stream into the original digital image with little loss in image quality. In this study, we assume the inputs of the JPEG decoding application are streams of JPEG images extracted from sample QCIF videos (174 × 144 resolution). Each JPEG image represents a frame of the video. The JPEG images are sequentially provided to the application to represent a multimedia streaming workload of a constant frame rate.

The program flow of the JPEG decoding application is shown in Fig. 6.10, which generally consists of two main parts. The first part is the initialization phase, in which the application builds up the data structure and reads the input image. Then in the second part, the application decodes the image by rows of Minimum Coded Units (MCUs). Each MCU row is a 174 × 16 fraction of the image.

6.3.1 Design-Time Scenario Clustering

First, we profiled the JPEG decoding workload with the inputs of 12 different QCIF video streams of various contents, to obtain the statistics of the execution time and energy of the workload. The number of JPEG images in each video ranges from 150 to 494. Then, the profiling traces are segmented into TNs of hundreds of microseconds. In this case, a TN is defined as the process of initialization or decoding an MCU row. Since one image is composed of nine MCU rows, the process of decoding one image is divided into ten TNs in this way.

Two parameters are selected as the monitoring parameters for the scenario detector. The first one is the different steps in the program flow (initialization,

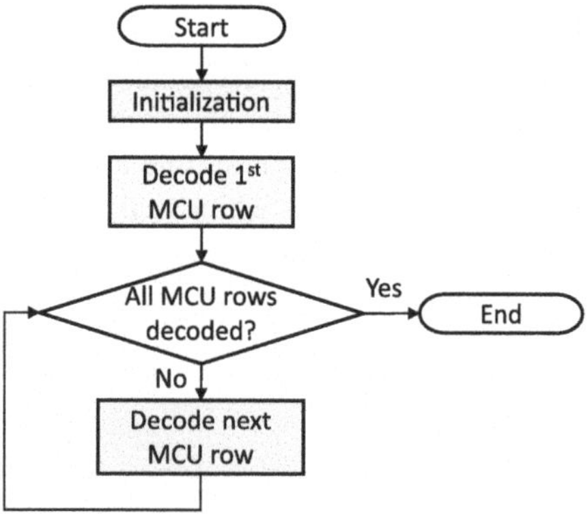

Fig. 6.10 Program flow of the JPEG decoding application. In the diagram, each step (gray square block) of the program flow is defined as a TN

decoding the first MCU row, decoding a middle MCU row, and decoding the last MCU row). Each of these steps exhibits distinct characteristics of execution time and energy consumption. Since the execution flow of the initialization step is the same for all images, its execution time and energy consumption are almost insensitive to the image content, which differentiates the characteristics of this step from the others. Then, in the main loop of decoding the image by MCU rows, the costs of processing the first and the last MCU rows are systematically different from the middle MCU rows. Therefore, the steps of processing the first and the last MCU rows are distinguished from the middle MCU rows to create their own scenarios.

The second monitoring parameter is the *bit per pixel* (BPP) of the image, which is defined as the length of the encoded bit stream divided by the number of pixels. BPP represents the data density of the image, which is related to the computation effort to decode an image. In the profiling, BPP is highly correlated to the execution time of decoding the MCU rows in the image. Also, the calculation of BPP is simple and can be done before reading the content of the encoded bit stream. Therefore, the controller can calculate the BPP of the upcoming TNs in the buffer to detect their scenarios. These properties make BPP a suitable monitoring parameter for scenario detection.

Figure 6.11 shows the relation between the number of instructions and the BPP for the TNs of decoding an MCU row. In general, the number of instructions increases with the BPP. Moreover, decoding the first MCU row takes more instructions, while decoding the last MCU row takes fewer instructions. In contrast, Fig. 6.12 shows that the execution time of the TNs of the initialization step is almost constant regardless of the input image content and the BPP. Based on these properties, nine dynamic scenarios are created, each corresponding to a certain step

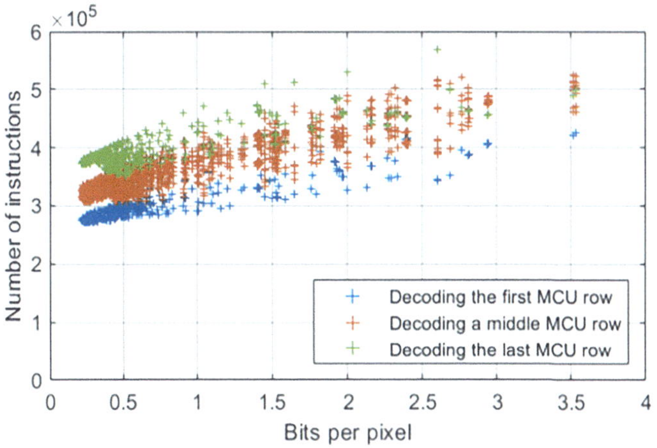

Fig. 6.11 Relation between the number of instructions and the BPP for the TNs of decoding an MCU row. It shows the high influence of BPP and MCU row location on the number of instructions

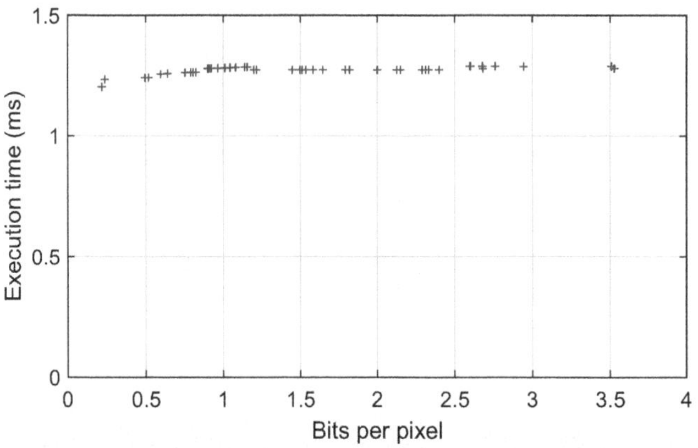

Fig. 6.12 Relation between the execution time under the 0.7 V/fast HDP mode and the BPP for the TNs of the initialization step. It shows the execution time is almost constant regardless of the input image content

Table 6.3 Scenario look-up table for the JPEG decoding application

Monitoring parameters			Cycle count	
Program flow	BPP range	Scenario	max	avg
Initialization	0–4.00	1	4771771	4703069
Decode the first MCU row	0–1.35	2	1423793	1271159
	1.35–4.00	3	1614701	1434089
Decode a middle MCU row	0–1.35	4	1609520	1414888
	1.35–2.10	5	1864506	1569669
	2.10–4.00	6	2175664	1767012
Decode the last MCU row	0 – 1.45	7	1838664	1661917
	1.45–2.50	8	2055572	1790937
	2.50–4.00	9	2266653	1991362

in the program flow and a BPP range. Based on the statistics of the cycle count of each scenario, the look-up table for scenario detection is created, which is shown in Table 6.3.

6.3.2 Run-Time Mitigation Results

At run-time, we selected an arbitrary QCIF video stream composed of ten JPEG images as the input of the workload. By applying the input, the performance variability is injected, in which the execution flow of processing each MPU row of each frame depends on the image content, so the execution time can be different and cannot be exactly calculated before execution. It requires 100 TNs for the JPEG decoders to process the input. For each TN encountered at run-time, the controller identifies which program flow step the TN is in and the BPP of the MCU row that the TN is processing. With these parameters, the controller can look up Table 6.3 to obtain the scenario of each TN and its average cycle count and upper bound cycle count, which is used by the controller algorithm (Algorithm 5.4) to derive the suitable modes to run each TN in the workload.

Figure 6.13 shows the DVFS and HDP modes selected during the entire run under the 8.0-ms frame period condition, and Fig. 6.14 shows the traces of slack of all approaches. The behaviors of each approach in these figures are similar to Sect. 6.2.2, though the workload is different. This shows that the characteristics of the approaches we discussed in Sect. 6.2.2 are not specific to a certain application. They are universal across workloads. Under a different workload, the multi-timescale DS-based approach can still run in a smooth slack curve which is always positive and close to zero. This proves that the merits of this approach are general and independent of workload.

Figure 6.15 shows the energy consumption of different approaches under different frame periods. Compared to the conventional single-DVFS WCET-based approach, the multi-timescale DS-based approach can save 16% energy on average and 25% energy on maximum. And compared to the DVFS BE approach, the energy consumption of the multi-timescale DS-based approach is about the same and 5% greater in the worst case. In some cases, the multi-timescale DS-based approach consumes even less energy, which is 6% in the best case. This result does not mean that the proposed approach can ensure timing guarantee without an energy penalty, since the energy consumption can be affected by multiple factors, such as the energy efficiency of the selected modes and the activities in the workload. Nevertheless, it shows that in some contexts the energy penalty can be minimal. Also, the energy differences among the approaches are smaller than in Case I in Sect. 6.2.2, which indicates that the JPEG decoding workload is less dynamic, so the potential of energy saving by mitigation is smaller, further showing that the energy saving achieved by mitigation is highly workload dependent.

6.4 Discussion of the Case Studies

We summarize the mitigation results of both benchmarks under all frame periods in Table 6.4, in which the average energy ratio is calculated with respect to the energy of the DVFS BE-based approach. The differences in the results of both benchmarks may be attributed to several factors. First, the degree of dynamism of the workload could affect the performance of mitigation approaches. A workload that is more dynamic in execution time, like Case I, could provide more potential for mitigation benefits. This explains why the energy consumption differences of methodologies are more prominent in Case I. Second, for the BE algorithm, the deadline miss rate varies between benchmarks, depending on the differences between the prediction and the real execution time. On the other hand, the WCET-based algorithm and the DS algorithm can both achieve zero deadline misses regardless of the benchmark.

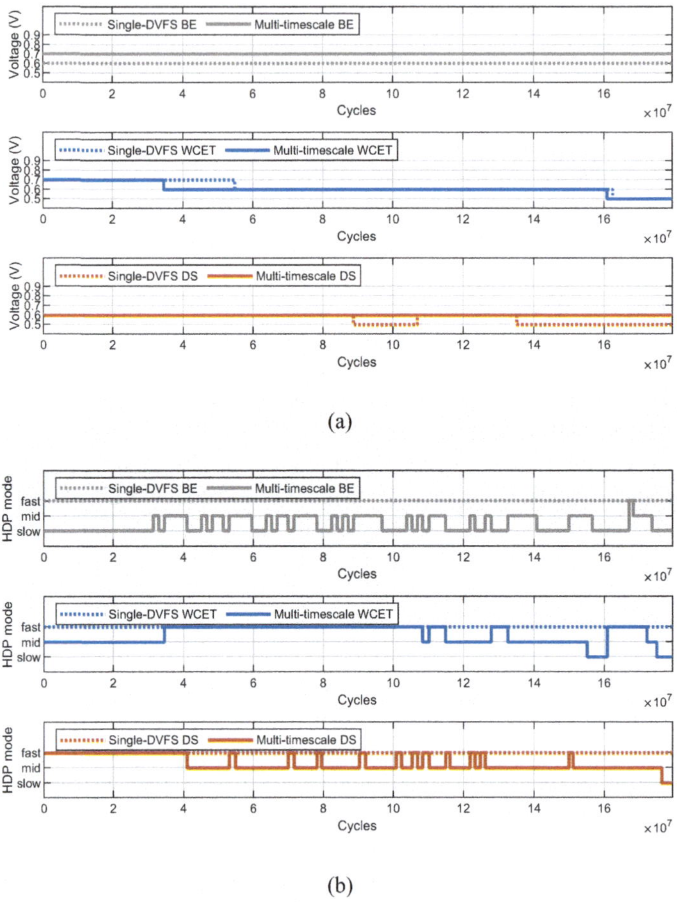

Fig. 6.13 Selected modes over the entire run of the JPEG decoding workload with an 8.0-ms frame period. (**a**) DVFS mode and (**b**) HDP mode

6.5 Summary

We have compared our proposed approach with conventional BE- and WCET-based approaches with two representative workloads (quantization of ADPCM encoding and JPEG decoding) running on the Ariane core. The comparisons between the cases with and without HDP have also been conducted. We tested these approaches on a cycle-accurate simulation flow in which the processor is simulated at the gate level, while the fast variability down to the picosecond level can be captured during the course of simulation. For both workloads we have demonstrated how dynamic scenarios can be created by program flow analysis and design-time profiling, followed by RTS-space segmentation. The run-time mitigation results show that when running a highly dynamic workload, the proposed approach can ensure

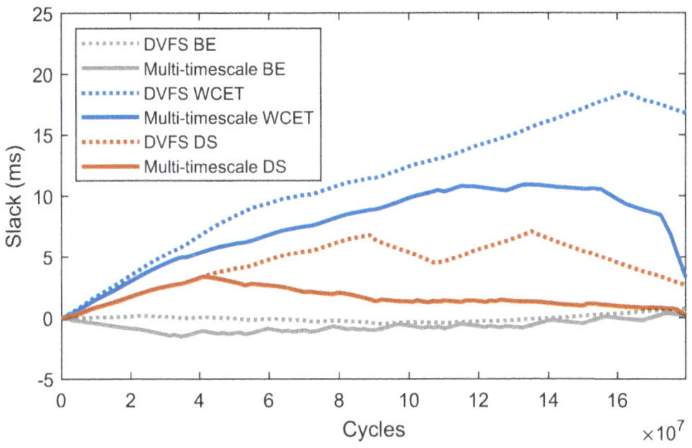

Fig. 6.14 Trace of slack over the entire run of the JPEG decoding workload with an 8.0-ms frame period

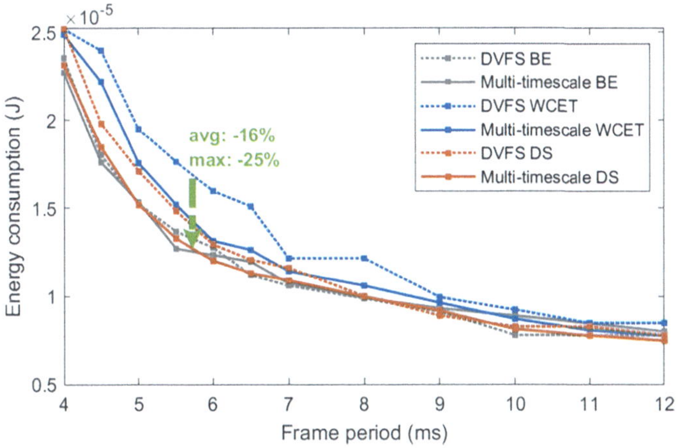

Fig. 6.15 JPEG decoding workload energy consumption of different methods under different time-criticality conditions

meeting all deadlines and at the same time reduce the dynamic energy consumption by 32% on average and 46% on maximum compared to the conventional WCET-based method with a single-DVFS knob. A deeper look into the results reveals that our approach can keep the workload progress ahead of the optimal schedule by only the minimal slack required to ensure no deadline misses. This can be attributed to both HDP and DS-based algorithm. The fast switching of HDP can switch for upcoming variability with the minimal latency, and DS-based algorithm can calculate the slack required for ensuring timing guarantees without pessimism. Therefore, our approach can ensure timing guarantees with minimal costs when compared to other approaches.

Table 6.4 Mitigation result summary with different methodologies and benchmarks

Approach	Case I: Quantization of ADPCM encoding		Case II: JPEG decoding	
	Deadline miss rate	Energy ratio[a] (min/avg/max)	Deadline miss rate	Energy ratio[a] (min/avg/max)
Single-DVFS BE	13%	0.47/0.59/0.95	52%	0.74/0.84/0.93
Multi-timescale BE	29%	0.49/0.58/0.92	46%	0.72/0.85/1.00
Single-DVFS WCET	0%	1.00/1.00/1.00	0%	1.00/1.00/1.00
Multi-timescale WCET	0%	0.72/0.90/1.00	0%	0.82/0.91/0.99
Single-DVFS DS	0%	0.66/0.83/1.00	0%	0.80/0.89/1.00
Multi-timescale DS	0%	0.54/0.68/0.97	0%	0.75/0.84/0.92

[a] Energy ratio is calculated w.r.t. the energy of the single-DVFS WCET method.

References

1. Falk, H., Altmeyer, S., Hellinckx, P., Lisper, B., Puffitsch, W., Rochange, C., Schoeberl, M., Sørensen, R.B., Wägemann, P., Wegener, S.: TACLeBench: a benchmark collection to support worst-case execution time research. In: Schoeberl, M. (ed.) 16th International Workshop on Worst-Case Execution Time Analysis (WCET 2016). OpenAccess Series in Informatics (OASIcs), vol. 55, pp. 2:1–2:10. Schloss Dagstuhl–Leibniz-Zentrum für Informatik, Dagstuhl (2016)
2. Guthaus, M., Ringenberg, J., Ernst, D., Austin, T., Mudge, T., Brown, R.: MiBench: a free, commercially representative embedded benchmark suite. In: Proceedings of the Fourth Annual IEEE International Workshop on Workload Characterization, pp. 3–14 (2001). https://doi.org/10.1109/WWC.2001.990739
3. Hu, Z., Buyuktosunoglu, A., Srinivasan, V., Zyuban, V., Jacobson, H., Bose, P.: Microarchitectural techniques for power gating of execution units. In: Proceedings of the 2004 International Symposium on Low Power Electronics and Design (IEEE Cat. No.04TH8758), pp. 32–37 (2004). https://doi.org/10.1145/1013235.1013249
4. Park, S., Park, J., Shin, D., Wang, Y., Xie, Q., Pedram, M., Chang, N.: Accurate modeling of the delay and energy overhead of dynamic voltage and frequency scaling in modern microprocessors. IEEE Trans. Comput. Aided Des. Integr. Circuits Syst. **32**(5), 695–708 (2013). https://doi.org/10.1109/TCAD.2012.2235126
5. Pering, T., Burd, T., Brodersen, R.: Voltage scheduling in the IpARM microprocessor system. In: ISLPED'00: Proceedings of the 2000 International Symposium on Low Power Electronics and Design, pp. 96–101 (2000). https://doi.org/10.1145/344166.344530
6. Pillai, P., Shin, K.G.: Real-time dynamic voltage scaling for low-power embedded operating systems. In: Proceedings of the Eighteenth ACM Symposium on Operating Systems Principles, SOSP '01, pp. 89–102. Association for Computing Machinery, New York (2001). https://doi.org/10.1145/502034.502044
7. Recommendation ITU-T G.722: 7 kHz audio-coding within 64 kbit/s. Tech. rep., ITU-T Test Signals for Telecommunication Systems (2012)

Chapter 7
Cross-Layer Reliability and Variability Mitigation Framework: Theory

Abstract In this chapter, we further discuss how to extend the state-of-the-art mitigation approaches to tackling device- and circuit-level performance variability. To achieve this, multiple mitigation approaches need to be integrated to form a cross-layer mitigation framework which can deal with different kinds of variability and errors. In the framework, each layer tackles a type of variability or errors produced from the lower layers and generates variability or errors which should be tackled by the upper layers. The framework can not only ensure timing guarantees but also ensure functional correctness at the same time. Here, the principles and theoretical basis of the framework are elaborated. Then, Chap. 8 will discuss the experiments on testing the capability, cost, and limitations of this mitigation framework.

This chapter is structured as follows: Sect. 7.1 describes the motivation and overview of the cross-layer framework, which consists of three layers. Then, we further elaborate on the details of each layer, respectively, from the bottom to the top in Sects. 7.2–7.4. The first layer, mitigation of device-level variability (process, voltage, and short-term aging effects), is described in Sect. 7.2. Then, the second layer, mitigation of errors, is described in Sect. 7.3. After that, we describe the last layer, mitigation of performance variability, in Sect. 7.4. Then, Sect. 7.5 describes how the overall reliability of the system can be evaluated with the cross-layer framework. Finally, we discuss some extensive questions about the framework in Sect. 7.6, including the comparisons with other cross-layer reliability frameworks and the applicability to other architectures and technologies.

7.1 Motivation and Principles

In Sect. 2.1, we have described variability sources which cause performance changes in devices. However, these variability sources do not directly cause changes in the execution time of workloads, since the performance of circuits is defined by the clock frequency, which is controlled by the phase lock loop, but not the intrinsic device performance. Nevertheless, the device-level variability is manifested by two effects. First, device-level variability causes timing violations, which causes errors

J.-Y. Lin et al., *Workload Dependent Mitigation Approaches for Performance Variability*, https://doi.org/10.1007/978-3-031-85624-2_7

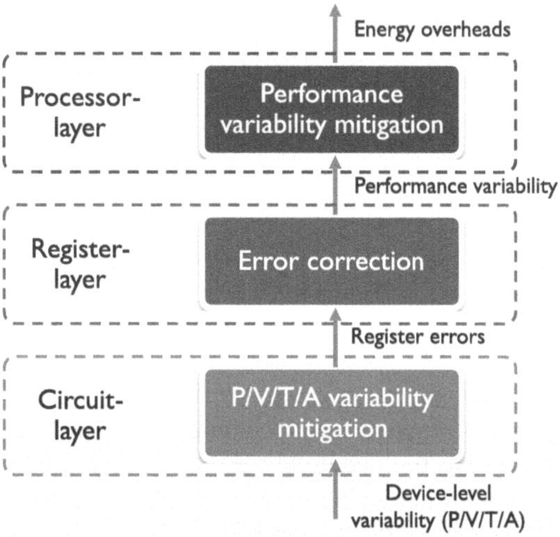

Fig. 7.1 Overall structure of the cross-layer mitigation framework

in registers. To ensure functional correctness, error correction mechanisms need to be applied, which induces execution time and energy overhead. Second, to reduce errors, modern processors have adopted mitigation approaches for device-level variability, as described in Sect. 2.4.2. The knobs used in these approaches, such as throttling and adaptive clocking, introduce extra execution time. Thus, mitigation for device-level performance variability is essentially compensating for the execution time overhead indirectly caused by the devices through these error correction and reduction mechanisms.

Therefore, we propose a cross-layer mitigation framework which integrates all of the mechanisms mentioned above. The framework can ensure functional correctness and timing guarantees in presence of the errors and performance variability caused by the devices. The overall structure, shown in Fig. 7.1, is composed of three layers. Each layer deals with different kinds of errors and variability at different levels.

First, in the circuit layer, run-time mitigation approaches are applied to reduce the variability of process, voltage, and short-term aging effects in devices. With these approaches, the performance fluctuation of devices is reduced, and so are timing violations and errors in registers. However, these mitigation approaches have their limits, so they cannot prevent all errors from happening. Therefore, there are still error occurrences which need to be tackled by error correction mechanisms in the upper layers. Moreover, these approaches cause performance fluctuation in the circuits, which also needs to be compensated by the performance variability mitigation approaches in the upper layers.

Second, in the register layer, error correction mechanisms fix the errors generated from the circuit layer, so functional correctness can be ensured. These mechanisms also generate execution time overhead, which contributes to performance variability. The performance variability needs to be tackled in the upper layer. The amount of time overhead increases with the number of errors required to be corrected. Hence, if we would leave out the device-level variability mitigation in the first layer, this layer could become overloaded with such overhead all the time, which is in practice too costly for the performance variability mitigation approach in the upper layer to deal with. Therefore, the first layer is essential to prevent this overloading of errors.

Last, in the processor layer, performance variability generated in the circuit and register layers is tackled by the performance variability mitigation approach, so timing guarantees can be ensured. Performance variability of the architecture level and the software level can also be tackled at the same time. The mitigation approaches operated at the cost of extra energy consumption, which is the final cost of the overall framework. The energy cost and the validity of timing guarantees depend on the amount and the frequency of performance variability. Only the fewer cases of performance variability needed to be tackled, the less energy cost and probability of failing timing guarantees. This shows the importance of the mitigation and error correction techniques in the lower two layers.

In short, the framework can ensure both functional correctness and timing guarantees in the presence of variability in device performance. The cost of the framework is only the extra energy consumption and the chip area required for the hardware of each mitigation approach.

7.2 Mitigation of Device-Level Variability

Mitigation approaches in the circuit layer aim to reduce the very fast device-level variability in process variation, voltage fluctuation, short-term aging effects, etc. The purpose is to keep the devices operating in a stable condition to minimize fluctuation in device performance. In this way, the circuit can work within its timing specification, so errors can be reduced. Though temperature fluctuation and long-term aging effects also cause variability in devices, they are normally tackled in higher layers due to their long time constants. In the following sections, we will discuss the state-of-the-art of these approaches and their limitations and overheads.

7.2.1 State-of-the-Art Mitigation Approaches

Voltage droop is the main target to deal with in this layer due to its nanosecond-level timescale, as described in Sect. 2.1.2. Voltage droop means occurrences of extraordinary drops of supply voltage that might cause errors in the circuit. It is normally caused by abrupt high demand for switching current from the devices that exceed the capability of the voltage regulation of the voltage supply module. As mentioned in Sect. 2.4.2, voltage droop mitigation approaches are proven to be effective in preventing voltage droop from producing errors.

As a representative example, here we demonstrate a typical voltage droop prevention technique. To prevent voltage droop, voltage sensors or critical path

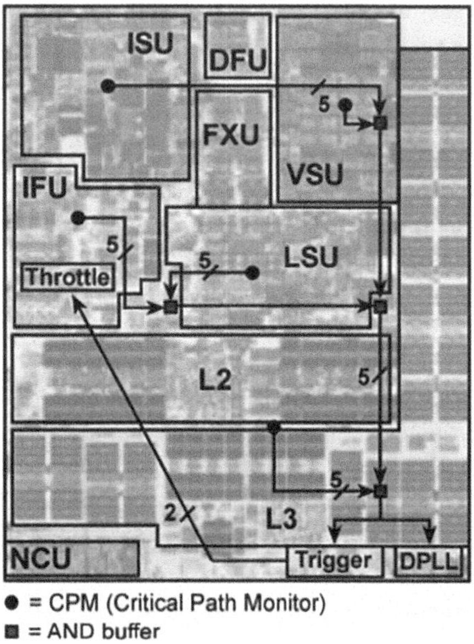

Fig. 7.2 Critical path monitor distribution in each core of IBM POWER7™ processor [18]. The AND buffers are used to aggregate the monitor signals, and the final signals could trigger knobs such as throttling or the digital phase-locked loop (DPLL)

monitors are distributed across the whole chip design to detect local events of large voltage decreases. Figure 7.2 shows an example of how critical path monitors are placed inside each core of IBM POWER7™ processor [18]. The readings of the sensors and monitors are passed to a dedicated controller to decide whether a potential voltage droop event is approaching. Once a potential voltage droop event is identified, the controller intervenes to cut the switching current demand in nanoseconds, by either throttling, adaptive clocking, or halting the processor cores. In this way, further supply voltage decrease is prevented, and so are errors in the circuit. As a result, Fig. 7.3 shows an example of measurements of a droop event with and without the voltage droop prevention approach. It clearly demonstrates the capability of the approach in cutting down voltage droop.

Instead of proactively eliminating variability, another concept is error-resilient design techniques [11] . In these techniques, in situ error detection circuits reactively detect errors in the circuit, which trigger on-site microarchitectural recovery mechanisms. Normally, the recovery process will cost one to several extra clock cycles, which is called "cycle noise." The designs of Razor [11, 12, 17] and tunable replica circuits (TRC) [8, 9] fall into this category. However, the extra circuitry leads to area and power overheads which could reach 6.9% and 9.4%, respectively [10]. The cycle noise also brings a performance loss that could reach 18% [19]. Apart from the overheads, a prominent drawback is that these techniques cannot guarantee that errors are detected under all circumstances, since the error detection is not realized by catching the wrong bits, but by sensing variability conditions which could generate errors, such as potential timing violations. Therefore, error-resilient designs are less popular than other mitigation approaches.

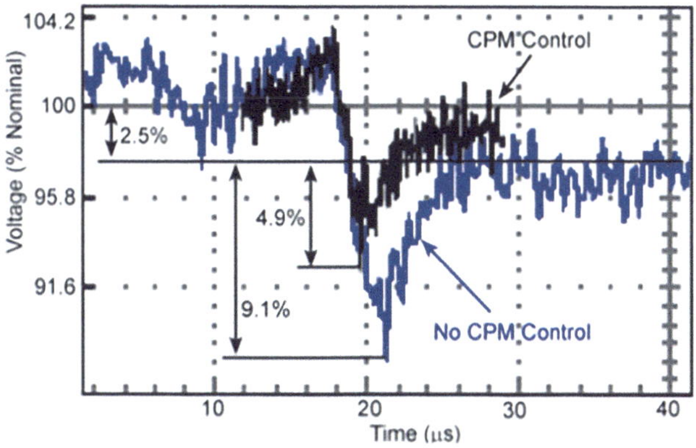

Fig. 7.3 Measurement of a droop event with and without the voltage droop prevention approach [18]

While mitigation approaches can effectively reduce variability to prevent errors, they have their limitations in the scale and the gradient of the variability they can tackle. As a result, guardbanding still needs to be used on top of the mitigation. However, due to the complexity of the underlying mechanisms, variability is not fully predictable, nor does it have a guaranteed bound. Therefore, the existing state-of-the-art approaches cannot guarantee to contain variability within their specifications and prevent all errors. It is still possible that a large and sharp variability exceeds the capability of mitigation and the guardband, causing errors in circuits. However, as motivated earlier in Sect. 7.1, the state-of-the-art circuit-level techniques proposed in the literature are still very useful and even essential to prevent the overload of errors in the register layer which will be discussed later in the next section.

7.2.2 Register-Level Errors

In the presence of variability, the change in device performance can exceed the timing specification of the circuit, causing timing violations. In consequence, registers in the circuit might catch incorrect bit values, which are called register-level errors. The error rate of the registers is highly related to the imposed margin of the guardbanding technique. In general, the relation between the error rate and the margin is depicted in Fig. 7.4, in which the larger the margin, the smaller the error rate. Due to recent trends of shrinking margins and increasing hardware complexity in advanced technologies, register-level errors are more and more frequent. Recent research in large-scale datacenters shows that silent data corruption induced by register-level errors is observed in one in thousands of processor cores [13, 24].

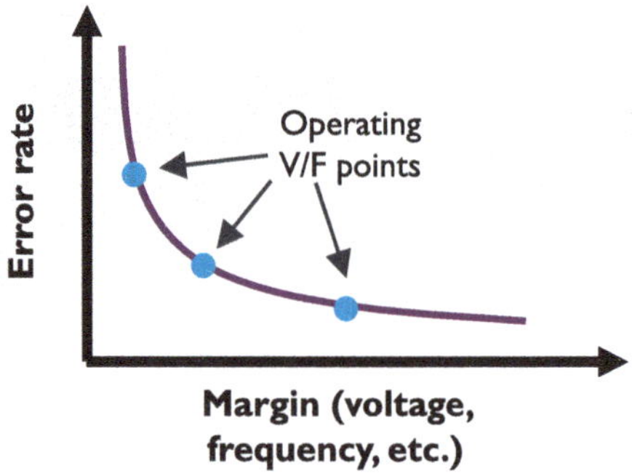

Fig. 7.4 Relation between the error rate and the margin

These errors are unpredictable and difficult to be prevented by testing in advance, so the costs of tackling them are very high.

The margin can be adjusted at run-time by changing the operating conditions, such as supply voltage and clock frequency. Ideally, the circuit is best to operate at the point of the smallest margin that does not cause any error. However, it is formidable to accurately calculate this ideal operating point since it depends on numerous factors, such as operating conditions, the amount of variability, workload, etc. Therefore, modern circuits normally operate at the point with a reasonable margin at which errors are rarely observed, and error correction mechanisms are applied to ensure functional correctness. The amount of margin affects the system costs in terms of speed, power, and area. On the one hand, a large margin means a high supply voltage or a slow clock frequency, which degrades circuit speed and power consumption. On the other hand, a small margin means more errors, which increases the time, energy, and circuit area invested in error correction. Therefore, the operating point of the circuit (supply voltage, clock frequency) has great influences on the system reliability (error rate, etc.) and system costs (speed, power, etc.) for ensuring reliability, so it should be carefully decided when designing the mitigation approaches for system reliability.

Estimation of the error rate is essential to evaluate the overall system costs of margining and mitigation approaches in the presence of register-level errors. However, very few studies have addressed the methods of error rate estimation in the past. Run-time measurements were performed in error-resilient design research to analyze the error rate dependency on voltage and frequency [8, 9, 12, 17]. Apart from measurements, some research has explored the idea of error rate evaluation by either design-time simulation or run-time estimation. In [26], the error rates detected by different types of monitors were compared in both simulation and experiments. In addition, the relation among the error rate of in situ monitors, voltage, and aging was simulated. In [6, 7], measurements were performed to find the relations between real circuit failures and error flags reported by critical path monitors, in order to minimize the voltage margin at run time. In [3], error probability is evaluated by gate-level simulation with degradation-aware cell libraries based on variations of voltage, temperature, and aging. In our work, we also simulate the error rate dependency on clock frequency later in Sect. 8.2.

In addition to register-level errors, mitigation approaches also bring performance variability. First, running the controller routine causes execution time overhead. Second, using system knobs to mitigate variability lengthens the execution time by increasing the number of clock cycles or reducing clock frequency. These execution time overheads need to be noted in the mitigation for performance variability in upper layers.

7.3 Register-Layer Error Correction

7.3.1 Error Correction Techniques

To ensure functional correctness, register-level errors generated in the circuit layer need to be tackled by error correction approaches. Many hardware (HW) and software (SW) approaches are developed in the past. However, most approaches either cannot provide full guarantees or come with significant timing, power, and energy overheads. Therefore, designing a highly reliable and low-cost error correction approach is still a challenging task. Time overhead is especially critical in time-critical and time-sensitive systems since it raises timing guarantee concerns. Hence, the time overhead needs to be compensated by the mitigation approaches for performance variability in the higher abstraction layers. However, if the error rate is too high, the time overhead could increase to unacceptable values, exceeding the capability of performance variability mitigation.

Many HW-based approaches exploit the principle of redundancy. This type of approach includes error-correcting codes used at different memory levels [2, 28, 33]. Other HW-based approaches are developed with dual modular redundancy (DMR) or triple modular redundancy (TMR) in different hardware granularities [14, 20, 31]. However, these approaches come with large timing, power, and area overhead, which is too costly for systems which are sensitive to either of these metrics. For instance, previous works show that ECC-protected L1 SRAM could have high overheads of 15% in area [35], 70% in energy [37], or more than 100% in time [37]. Also, a processor chip with full DMR shows more than two times the area of the original processor [5]. While these overheads can be reduced by using these techniques only on circuits of critical functions, reliability is also reduced. Another HW-based concept is error-resilient designs discussed in Sect. 7.2.1, which also suffer from the problems of high overhead and compromised reliability.

SW-based error correction approaches can be classified into techniques that apply checkpointing and rollback-recovery (CRR) or resource redundancy. In the former techniques, checkpointing is used to save the software states on a regular basis. In cases of an error, a recent checkpoint state is restored by recovery processes. While this technique has been used in plenty of works [1, 15, 34, 39], it introduces significant overheads that could reach 70% in energy and more than 100% in execution time [37]. In cases of high error rates, the recovery events can even overflood the system, so that the time overhead is prohibitive to be mitigated.

In the resource redundancy approaches, both temporal and spatial redundancy can be exploited, As for temporal redundancy, workloads can be duplicated in the same core before their results are used as the source operands of the follow-up instructions [27, 34]. As for spatial redundancy , workloads are duplicated in a nearby idle core [34]. The granularity of the duplication can vary from the instruction level [27] to the process level [34]. However, the operation of these techniques relies on spare resources such as idle time and cores, which are not always available. Therefore, their reliability cannot be fully guaranteed.

To sum up, most of the proposed error correction techniques face the dilemma of introducing significant overheads or degrading reliability. Recovery-based approaches tend to have high time overheads, while redundancy-based approaches tend to have high area and energy overheads. To reduce overheads, many approaches turn to partially protected circuits or data, which compromise reliability.

7.3.2 HW/SW Reliability Mitigation with OCEAN

To overcome the drawbacks and limitations mentioned in the last session, the concept of integrated HW/SW error correction has been proposed [21, 22]. Cross-layer error correction design can combine and optimize the error-resilient approaches in different abstraction layers, such as propagating information on reliability from the HW level to the SW level to reduce cost. Although cross-layer designs are promising to ensure reliability with minimized total overheads, their challenge lies in deciding the proper techniques and granularities which can globally optimize the interlayer communications and the holistic costs. Suboptimal solutions will lead to one optimized parameter (e.g., time) at the expense of another suboptimal parameter (e.g., energy).

To have a low-overhead error correction approach, our framework adopts OCEAN , the optimized HW/SW reliability mitigation approach [37], developed by Sabry et al. It provides a generalized framework for cross-layer optimization, achieving optimal error correction. A hybrid HW/SW mitigation approach is realized with this framework. In this approach, the HW reliability is enhanced by using an error-protected buffer, as shown in the schematic in Fig. 7.5. The buffer is utilized to protect a portion of the processed data used to restore from run-time errors. By optimizing the buffer size, energy overhead can be minimized under timing and area constraints. In addition, a hardware module (referred to by μDMA) resembling the direct memory access function is dedicated to the transfer between the memory module and the buffer, further removing the latency in data movement.

In the SW layer, OCEAN applies the checkpointing and rollback-recovery (CRR) technique. An example is shown in Fig. 7.6. In this approach, each workload is segmented into computation phases. The collection of computation results generated in each computation phase is called *data chunk*. At the end of each computation

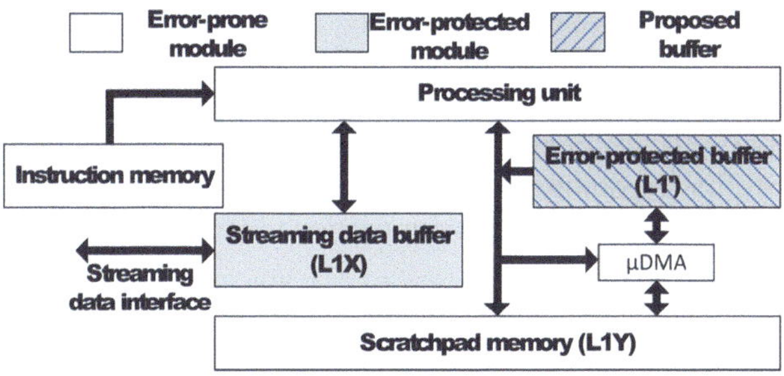

Fig. 7.5 Schematic diagram of OCEAN, the optimized HW/SW reliability mitigation approach [37]

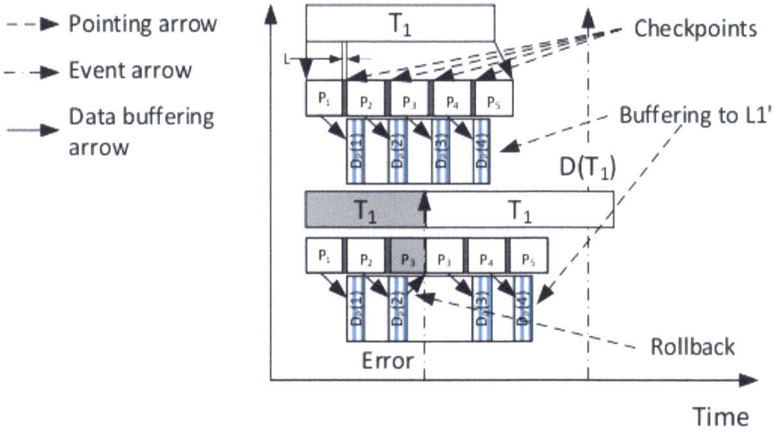

Fig. 7.6 A CRR example of dividing a task T1 to five phases (P1–P5), showing the impact on error mitigation. The faulty task/phase is shaded with light gray, the checkpoints are shaded with dark gray, and the data chunks are shaded in stripe-patterned blue [37]

phase, a checkpoint is inserted to check the errors in the data chunk with error detection techniques such as signature comparison. If no error is detected, the data chunk is considered error-free, and it is copied to the error-protected buffer as a backup by the μDMA. Meanwhile, the main program can proceed to the next computation phase. The main program can run in parallel with the copy of the data chunk, so the time overhead is minimized.

On the other hand, if errors are detected by the checkpoint routine, rollback-recovery occurs, as illustrated in P3 of Fig. 7.6. First, a rollback routine is performed, in which the main program is recovered to the state of the start of the current phase, and the data chunk of the last phase is recovered by copying back from the error-protected buffer. Then, the main program reruns the current phase. At the end of this phase, a new checkpoint is inserted to check the errors in the recomputed data chunk. The phase can roll back and recompute repeatedly until the data chunk is proven correct by the checkpoint routines. Therefore, there is no bound for the number of rollbacks required for each phase.

In experimental results, OCEAN achieves full error mitigation with 10.1% average energy overhead compared to the baseline operation that does not include any error correction capability and 65% energy saving compared to a cross-layer error mitigation mechanism without optimization [37].

7.3.3 Theoretical Analysis of Time Overhead

Our work further extends the use of OCEAN to be integrated into the cross-layer reliability and variability mitigation framework. In order to have a global view of the impact of the time overhead of error correction on timing guarantees, it is crucial to analyze the interplay between errors and timing guarantees. Therefore, we derive the relations between errors and time overhead under the CRR mechanism. Moreover, the limit within which the system can ensure reliability can be estimated by this analysis. Our analysis in this section is workload independent, assuming a static error rate and a constant size of each computation phase. We will further perform workload-dependent analysis with variable segment sizes in Sect. 8.3.

In our error model, a cycle is erroneous if any register of a pipeline stage contains a wrong value. The probability that a cycle is erroneous is static over time, which represents the effect of variability in logic circuits during the execution of the application. Assuming a certain segment of the workload is concerned, the probability that there is no error during this segment is as follows:

$$Pr(N_e{=}0) = (1-p)^{C_{\mathrm{seg}}} \tag{7.1}$$

where N_e is the number of erroneous cycles, C_{seg} is the total number of cycles in this segment, and p is the probability of a cycle to be erroneous, respectively. The number of rollbacks for each segment follows Geometric distribution, derived with Eq. 7.1, as follows:

$$Pr(N_{\mathrm{rb}}{=}n_{\mathrm{rb}}) = (1-(1-p)^{C_{\mathrm{seg}}})^{n_{\mathrm{rb}}}\,(1-p)^{C_{\mathrm{seg}}} \tag{7.2}$$

Figure 7.7 shows the probability distribution of the number of rollbacks of a segment, assuming the error probability of the segment to be 0.2. The distribution shows that multiple rollbacks are possible but highly unlikely. Moreover, by the probability distribution of Eq. 7.2, we can derive the expected number of rollbacks as follows:

$$E[N_{\mathrm{rb}}] = \frac{1-(1-p)^{C_{\mathrm{seg}}}}{(1-p)^{C_{\mathrm{seg}}}} \tag{7.3}$$

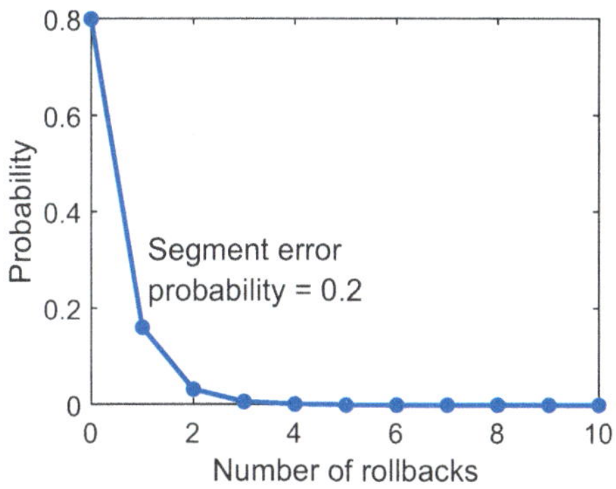

Fig. 7.7 Probability distribution of the number of rollbacks of a segment, assuming the error probability of the segment is 0.2

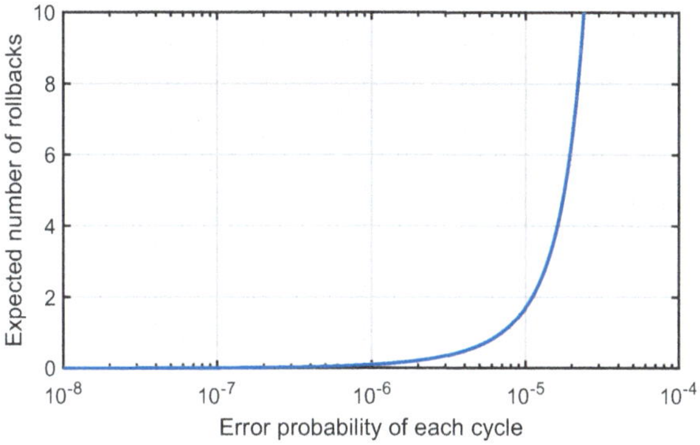

Fig. 7.8 Expected number of rollbacks for a segment of 10^5 cycles, under different error probability of cycles

Figure 7.8 shows the relation between the expected number of rollbacks and the error probability of each cycle, assuming $C_{\text{seg}} = 10^5$. The figure indicates that in a small error probability window, the number of rollbacks changes rapidly from almost zero to more than ten. Therefore, the number of rollbacks is highly sensitive to the error rate, and with the exponential growth in error rate, the number of rollbacks will quickly increase to unacceptable values.

The time overhead of CRR can also be evaluated with the error model. Here we assume a time model in which a checkpoint routine is executed at the end of each segment, and a rollback routine is executed at the start of each rollback-recovery occurrence. Therefore, we can derive the time overhead of each segment by cycles as follows:

$$C_{\text{ov}} = C_{\text{cp}} + n_{\text{rb}}(C_{\text{rb}} + C_{\text{seg}} + C_{\text{cp}}) \tag{7.4}$$

where n_{rb} is the number of rollbacks, and C_{ov}, C_{cp}, and C_{rb} are the number of cycles of the time overhead, checkpoint routine, and rollback routine, respectively. Here, we assume a CRR system model by abstracting the OCEAN approach described in Sect. 7.3.2, in which a checkpoint routine contains 100 cycles and a rollback routine contains 48 cycles. Then, combined with Eqs. 7.3 and 7.4, the expected value of

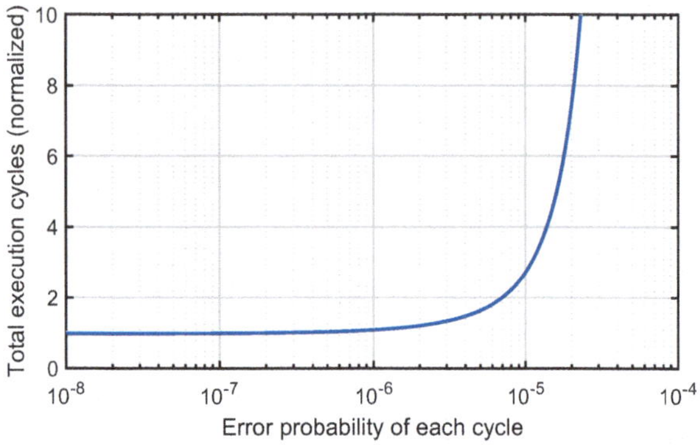

Fig. 7.9 The expected value of the total execution cycles normalized to the original execution cycles for the segment (10^5 cycles), under different error probability of each cycle

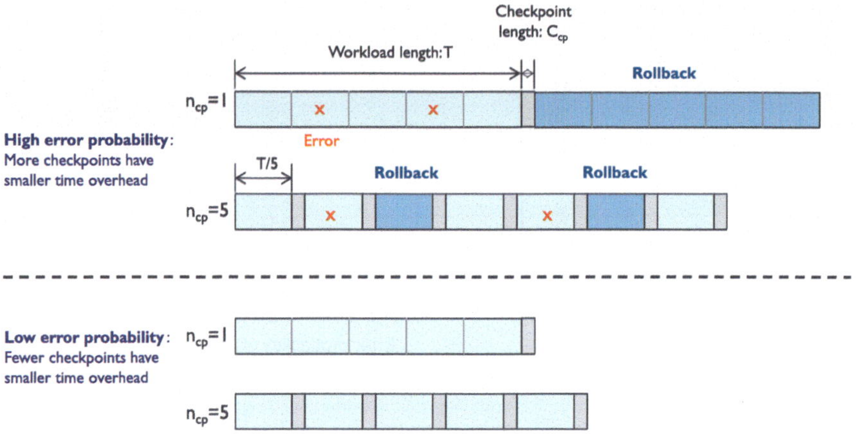

Fig. 7.10 Example of how the number of checkpoints affects the time overhead under different error probability conditions

time overhead can be calculated in cycles. Figure 7.9 shows the relation between the expected value of the total execution cycles and the error probability of each cycle, assuming $C_{seg} = 10^5$. Similar to the number of rollbacks shown in Fig. 7.8, time overhead quickly increases to unfeasible values with the exponential growth in error rate.

7.3.4 *Optimal Checkpointing*

In CRR schemes, the amount of time overhead is significantly affected by the number of checkpoints and rollbacks performed. However, there is a trade-off between the time overhead generated from checkpoints and rollbacks. As shown in the example of Fig. 7.10, more checkpoints increase the total time spent in checkpointing routines. Nevertheless, more checkpoints also mean smaller segment

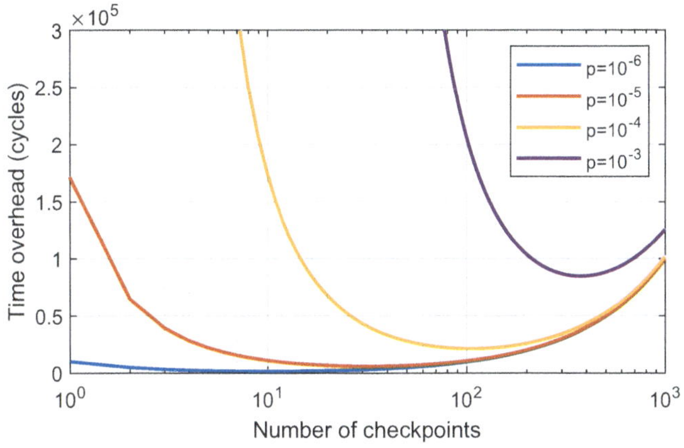

Fig. 7.11 Relations between time overhead and the number of checkpoints for a segment of 10^5 cycles, under different error probability of cycles (p)

sizes, which reduce the rollback and recovery time spent in each error occurrence and thus reduce the total time spent in rollback-recovery.

Figure 7.11 shows the relations between time overhead and the number of checkpoints for a segment of 10^5 cycles. For different error probabilities of cycles (p), there is an optimal number of checkpoints which can minimize total time overhead. In general, more checkpoints are favored at high error probability, since the time overhead is dominated by the rollback-recovery time. On the other hand, at low error probability, time overhead is dominated by checkpointing routines, so fewer checkpoints are preferred to reduce the total time spent in checkpointing routines.

If an ideal CRR scheme is assumed, in which checkpoints can be inserted anywhere in the workload and the lengths of checkpoint routines are constant, then there is an optimal number of checkpoints that could minimize the expected value of total time overhead. Based on the derivation in [40], the optimal number of checkpoints is calculated as follows:

$$n_{\mathrm{cp,opt}} = -\frac{1}{2}\ln P + \sqrt{\frac{1}{4}(\ln P)^2 - \frac{C_{\mathrm{wl}}\ln P}{C_{\mathrm{cp}}}} \tag{7.5}$$

where $n_{\mathrm{cp,opt}}$ is the optimal number of checkpoints. C_{wl} and C_{cp} are the number of cycles of the workload and the checkpoint routine, respectively. P is the probability that there is no error in every cycle of the workload, which can be calculated as follows:

$$P = (1-p)^{C_{\mathrm{wl}}} \tag{7.6}$$

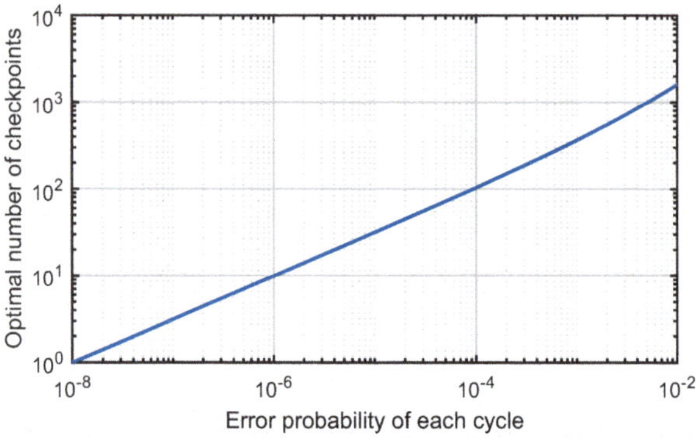

Fig. 7.12 Optimal number of checkpoints under different error probabilities of each cycle

Fig. 7.13 Expected value of total execution cycles with the optimal number of checkpoints compared to the scheme of one checkpoint per segment, under different error probabilities of each cycle

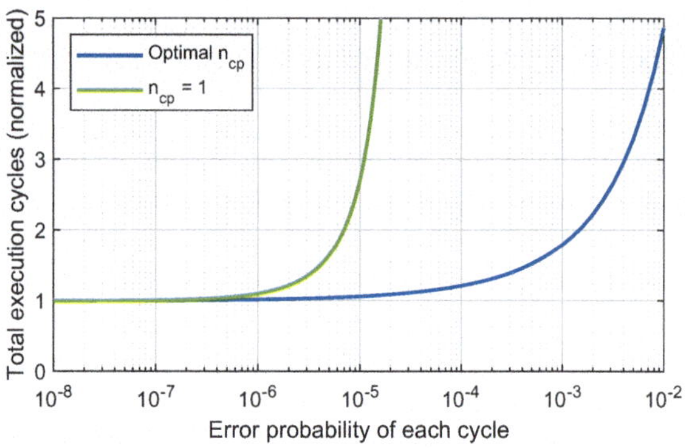

where p is the probability of a cycle being erroneous. The time required for the rollback routine is not considered in the calculation. Nevertheless, the rollback routine in our CRR scheme is relatively short (48 cycles) compared to the workload length, so the difference in the calculation is negligible. Figure 7.12 shows the optimal number of checkpoints under different error probabilities of each cycle. It indicates that the optimal number of checkpoints increases with the growth of error probability.

Figure 7.13 shows the minimized time overhead with the optimal number of checkpoints, compared to the scheme of one checkpoint per segment in Fig. 7.9. With variable numbers of checkpoints, the total execution cycles increase by a slower rate with respect to the error probability. Moreover, the optimal checkpointing scheme reaches a significantly higher error probability before the surge of the total execution cycles, showing much better system reliability.

7.4 Processor-Layer Mitigation of Performance Variability

In the processor layer, mitigation approaches are used to tackle the performance variability to ensure timing guarantees. In Chaps. 5 and 6, we have presented the workload-dependent multi-timescale mitigation approach to deal with performance variability of the architecture and software levels. Here, the use of the approach is extended to include performance variability from the lower levels, such as the time overhead of the CRR mechanism. As we will demonstrate in Sect. 8.3, by enlarging the time budget of each TN, the mitigation approach can absorb the time overhead of the CRR mechanism to prevent deadline misses.

In general, the time budget assigned to each TN in the adaptive control algorithms affects its capability of ensuring timing guarantees in the presence of register-level errors. Conservative algorithms supply TNs with larger budgets, which could absorb the time overhead. Therefore, conservative algorithms are better at meeting deadlines at high error probability. However, the high processor speed used by conservative algorithms consumes more energy, which is unnecessary if errors are rare. On the other hand, aggressive algorithms assign smaller budgets to each TN, which might fall short of meeting deadlines. However, it is better at energy saving. Therefore, it is important to schedule TNs with the right amount of time budgets which can ensure timing guarantees and save energy.

Furthermore, since errors occur randomly, there is no mathematical bound of the time overhead, which makes the timing guarantees impossible to be fully ensured. However, conditions of extremely large time overhead, like multiple errors in a very short time, are not realistic in the real world. Therefore, the number of errors can be bound by realistic specifications. Also, in robust system designs, fail-safe mechanisms should be applied on top of the mitigation approaches to prevent severe consequences and recover to normal states under extreme conditions. Thus, functional correctness and timing guarantees can still be fully ensured.

7.5 System Reliability Evaluation

While effective in handling variability and errors, every mitigation approach has its own limits. If the amount of variability and errors exceeds the capability of the mitigation approach, it could fail the guarantees it promises and lead to the loss of system reliability. In the mitigation of device-level variability, an excessively large and sharp voltage droop could make the voltage droop prevention approach unable to raise the supply voltage to a sufficient voltage level, causing circuit failure. In the error correction approaches, too many errors could overwhelm the error correction mechanism, so it cannot fix all the errors in a reasonable amount of time. In performance variability mitigation, the capability of absorbing the time overhead is constrained by the maximum processor speed. Excessive execution time overhead could make the mitigation approach unable to offset the performance variability even with the highest processor speed, causing deadline misses.

Therefore, the reliability of the whole system is constrained by the limit of each mitigation approach. To evaluate the overall system reliability, it is required to analyze the boundary conditions of the mitigation approaches and interdependencies among different approaches under these conditions. With such an analysis, we discovered that the system reliability of the presented framework can only be ensured within a limit of the error probability per cycle, which we refer to as the "error rate wall" (ERW). ERW can be calculated by analyzing the interdependencies of the error correction and the performance variability mitigation, which we will describe as follows.

Figure 7.14 depicts a basic example of performance variability mitigation dealing with a simple workload with one deadline and an amount of execution cycle overhead. Suppose the workload has C_{wl} cycles originally, C_{ov} cycles of overhead, and C_{tot} cycles in total including the overhead. We assume $C_{\max}$ is the maximum number of cycles the processor can run with the highest speed between the start of the workload and the deadline, which represents the maximum available computing resource of this interval. Whether the workload with overhead can finish before the deadline depends on the relation of C_{tot} and $C_{\max}$. When C_{ov} is small so that $C_{\mathrm{tot}} < C_{\max}$, the execution time overhead can be offset by raising the processor speed. If C_{ov} increases so that $C_{\mathrm{tot}} = C_{\max}$, then the processor is just fast enough to finish the workload at the deadline with maximum processor speed. If C_{ov} further increases so that $C_{\mathrm{tot}} > C_{\max}$, the total execution cycles are so large that even the maximum processor speed cannot save the workload from violating the deadline. Therefore, it requires $C_{\max} \geq C_{\mathrm{tot}}$ for the performance variability mitigation to ensure timing guarantees.

The above relation shows that the sizes of the original workload and the overhead both affect the viability of performance variability mitigation. The ratio of C_{wl} to $C_{\max}$ is known as the *utilization* (U) of the workload, representing the proportion of resource the original workload occupies out of the total computing resource. Therefore, the aforementioned criterion can be expressed by the ratio of C_{tot} and C_{wl} as follows:

$$\frac{C_{\mathrm{tot}}}{C_{\mathrm{wl}}} \leq \frac{1}{U} \tag{7.7}$$

Equivalently, the relation can be expressed with C_{ov} and C_{wl}:

$$\frac{C_{\mathrm{ov}}}{C_{\mathrm{wl}}} \leq \frac{1-U}{U} \tag{7.8}$$

The ratio of C_{ov} to C_{wl} represents the allowed amount of the overhead compared to the original workload which can still ensure timing guarantees. The allowed overhead also depends on the utilization of the original workload.

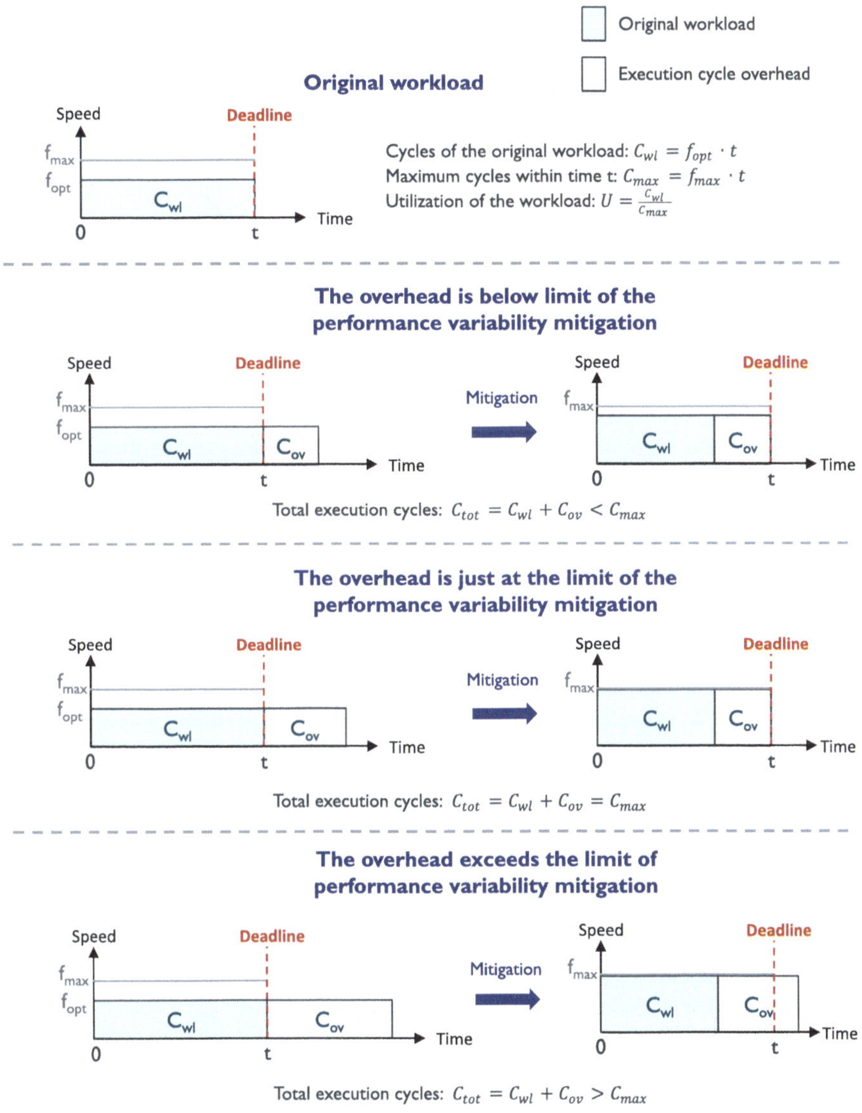

Fig. 7.14 Illustration of the limit of the performance variability mitigation in mitigating the execution cycle overhead of a workload

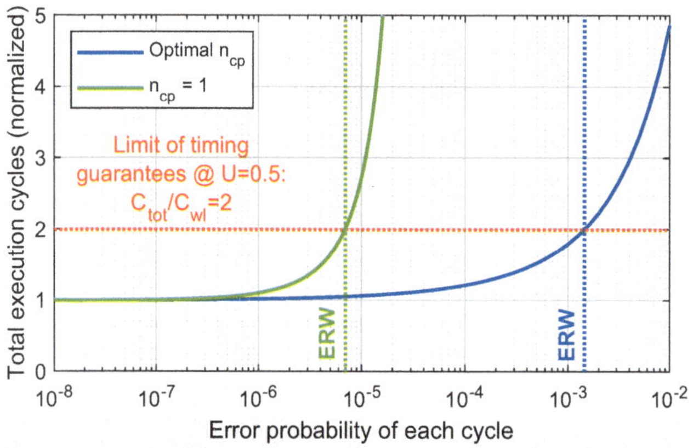

Fig. 7.15 Error rate walls under 0.5 utilization of the original workload in the CRR schemes of Fig. 7.13

In our framework, the execution cycle overhead is generated by the CRR, in which the execution cycles are dependent on the error probability of each cycle, as shown in Fig. 7.13. Therefore, combining the above equation with the relation between the total execution cycles and error probability, we can estimate the error rate limit within which the timing guarantees can be ensured, namely the ERW. Beyond the ERW, the reliability of functional correctness and timing guarantees can no longer be ensured simultaneously.

For example, assuming the CRR schemes of Fig. 7.13 with the utilization of the original workload is 0.5, by Eq. 7.8 we derive that the $C_{\text{tot}}/C_{\text{wl}}$ is 2. Under these conditions, the ERW can be obtained by finding the crossing point of the curve with the y-axis level of 2, yielding approximately 7×10^{-6} for the optimal checkpointing scheme and 1.4×10^{-3} for the scheme of one checkpoint per segment, as shown in Fig. 7.15. The ERW is improved by more than 200 times for the optimal checkpoint compared to the case of one checkpoint per segment, which shows the potential of enhancing reliability by optimization of checkpointing.

While the utilization can change over time, the highest utilization during the entire workload should be used in the calculation, since it is the bottleneck of timing guarantees. With this analysis, we can define the feasible region of the system specification in which reliability can be ensured. The validity of this analysis can be further proven by simulation with real workload demonstrated later in Chap. 8.

7.6 Discussion

7.6.1 Comparison with Other Cross-Layer Mitigation Approaches

In addition to our framework, cross-layer approaches have been researched by others in various layers for different reliability constraints. These approaches mainly aim to optimize for multiple goals (i.e., minimizing energy) under certain reliability constraints. Though distinct in the applied mitigation techniques, most of these approaches are complementary to our framework in that they are deployed in higher layers, such as the OS and application layers [23].

In the error correction field , many works have explored the benefits of cross-layer approaches [21, 22], apart from the OCEAN approach applied in our framework. For example, Lee et al. [30] present a cross-layer soft error mitigation approach for multimedia, which combines error detection codes in hardware, drop and forward recovery in middleware, and error-resilient video encoding at the application level. Moreover, Henkel et al. [23] proposed a multilayer dependency framework of functional correctness spanning from the microarchitecture layer to the application layer. These approaches operate on similar principles but in distinct layers and application fields from our framework, so they can be combined with our framework to cover more abstraction layers and fields.

In real-time scheduling fields, plenty of work addresses the problem of scheduling in the presence of recovery events due to hardware errors [16, 29, 32, 38, 41]. However, these works often assume oversimplified mathematical models of hardware, error occurrences, and CRR mechanisms, which make them inadequate to reflect real-world scenarios. For example, the assumptions of the error model can be a single error [32], a bounded number of errors [29], or Poisson distribution [41], and they can assume the recovery processes to be free of errors [32]. These limitations result in discrepancies between the model and the real statistics of error occurrences. Hence, our framework can complement these insufficiencies by providing realistic hardware and error models. Regardless of the aforementioned flaws, these approaches can serve as multitask scheduling solutions which work in a higher layer on top of our framework.

In addition to functional correctness and timing guarantees, many approaches also address other reliability objectives (i.e., thermal constraints and lifetime), as discussed in Sect. 2.4.2. These techniques are also included in cross-layer approaches of multiple reliability objectives. For example, Ansari et al. [4] proposed an adaptive CRR approach under thermal design power (TDP) constraints. In addition, previous works have proposed real-time scheduling approaches while considering thermal constraints [36] or lifetime [25] at the same time. These approaches can be complementary to our work in addressing more reliability requirements, completing the framework with the full reliability scope.

7.6.2 Applicability to Other Architectures and Technologies

Although the presented framework is instantiated for a single-core processor, the principles of this framework can be applied to other architectures and technologies. The first layer, the mitigation of device-level variability, can be any domain-specific mitigation approaches which target the error-generating mechanisms prominent for that technology and domain. For example, in a space application, the Single Event Upset (SEU) caused by radiation could be the dominant source of errors, which is mainly mitigated by the radiation-hardened circuits to reduce the error probability of the system. The second layer, the register-layer error correction, could also be realized by alternative error correction techniques which are cost-effective for that architecture and application. For example, the error correction codes (ECCs) in the communication domain can also be used for checking the register errors. Finally, for the last layer, the applicability of the performance variability to the other architectures and technologies has been elaborated in Sect. 5.6.

Furthermore, although in the presented framework only the functional correctness and timing guarantees are targeted, the framework is extendable to include other reliability requirements and QoR constraints, such as chip area and thermal constraints. In this case, the condition of ensuring guarantees is modeled as a constraint function in an N-dimensional Pareto space. While the concepts of dynamic scenarios are still applicable, the processes of scenario clustering and run-time controlling algorithm should be adjusted to consider more constraints. Moreover, the interdependencies of these constraints should be carefully analyzed to identify the boundaries of guarantees.

HDP knob is now only demonstrated on the processor datapath, but the principles are reusable also for communication networks and for the memory organization (as we discussed before). Moreover, also DVFS is at least partly reusable for the memory organization, especially for SRAM-based memories.

In summary, it is feasible to adapt and extend the mitigation framework based on the same mitigation principles for other technologies, architectures, applications, domains, and QoS constraints. Therefore, the system designers should find the most cost-effective combination of mitigation approaches for the targeted system to implement the cross-layer mitigation framework on the same principles.

7.7 Summary

We propose a cross-layer mitigation framework to tackle the device-level and circuit-level performance variability. The framework can ensure both functional correctness and timing guarantees by three mitigation approaches in a layered structure. In the circuit layer, mitigation approaches decrease device-level variability (process, voltage, temperature, and aging) to reduce register-level errors. Then, in the register layer, error correction mechanisms such as CRR fix register-level errors at the cost of time overhead. Last, in the processor layer, mitigation approaches ensure timing guarantees in the presence of performance variability generated from the lower layers. Our framework is extendable with other mitigation approaches in different layers, achieving a full reliability framework across all layers and objectives.

References

1. Abate, F., Sterpone, L., Violante, M.: A new mitigation approach for soft errors in embedded processors. In: 2007 9th European Conference on Radiation and Its Effects on Components and Systems, pp. 1–6 (2007). https://doi.org/10.1109/RADECS.2007.5205504
2. AMD: AMD eighth-generation processor architecture. Tech. rep., AMD white paper (2001). http://intel80386.com/amd/k8_architecture.pdf
3. Amrouch, H., Ehsani, S.B., Gerstlauer, A., Henkel, J.: On the efficiency of voltage overscaling under temperature and aging effects. IEEE Trans. Comput. **68**(11), 1647–1662 (2019). https://doi.org/10.1109/TC.2019.2916869
4. Ansari, M., Safari, S., Khdr, H., Gohari-Nazari, P., Henkel, J., Ejlali, A., Hessabi, S.: Power-aware checkpointing for multicore embedded systems. IEEE Trans. Parallel Distrib. Syst. **33**(12), 4410–4424 (2022). https://doi.org/10.1109/TPDS.2022.3188568
5. Bannon, P., Venkataramanan, G., Sarma, D.D., Talpes, E.: Computer and redundancy solution for the full self-driving computer. In: 2019 IEEE Hot Chips 31 Symposium (HCS), pp. 1–22 (2019). https://doi.org/10.1109/HOTCHIPS.2019.8875645
6. Benhassain, A., Cacho, F., Huard, V., Mhira, S., Anghel, L., Parthasarathy, C., Jain, A., Sivadasan, A.: Robustness of timing in-situ monitors for AVS management. In: 2016 IEEE International Reliability Physics Symposium (IRPS), pp. CR-4-1–CR-4-6 (2016). https://doi.org/10.1109/IRPS.2016.7574593
7. Benhassain, A., Mhira, S., Cacho, F., Huard, V., Anghel, L.: In-situ slack monitors: taking up the challenge of on-die monitoring of variability and reliability. In: 2016 1st IEEE International Verification and Security Workshop (IVSW), pp. 1–5 (2016). https://doi.org/10.1109/IVSW.2016.7566606
8. Bowman, K., Tschanz, J., Wilkerson, C., Lu, S.L., Karnik, T., De, V., Borkar, S.: Circuit techniques for dynamic variation tolerance. In: 2009 46th ACM/IEEE Design Automation Conference, pp. 4–7 (2009). https://doi.org/10.1145/1629911.1629915
9. Bowman, K.A., Tschanz, J.W., Lu, S.L.L., Aseron, P.A., Khellah, M.M., Raychowdhury, A., Geuskens, B.M., Tokunaga, C., Wilkerson, C.B., Karnik, T., De, V.K.: a 45 nm resilient microprocessor core for dynamic variation tolerance. IEEE J. Solid-State Circuits **46**(1), 194–208 (2011). https://doi.org/10.1109/JSSC.2010.2089657
10. Bull, D., Das, S., Shivashankar, K., Dasika, G.S., Flautner, K., Blaauw, D.: A power-efficient 32 bit ARM processor using timing-error detection and correction for transient-error tolerance and adaptation to PVT variation. IEEE J. Solid-State Circuits **46**(1), 18–31 (2011). https://doi.org/10.1109/JSSC.2010.2079410
11. Das, S., Bull, D., Whatmough, P.: Error-resilient design techniques for reliable and dependable computing. In: 2014 IEEE International Reliability Physics Symposium, pp. 3D.2.1–3D.2.6 (2014). https://doi.org/10.1109/IRPS.2014.6860627
12. Das, S., Tokunaga, C., Pant, S., Ma, W.H., Kalaiselvan, S., Lai, K., Bull, D.M., Blaauw, D.T.: RazorII: In situ error detection and correction for PVT and SER tolerance. IEEE J. Solid-State Circuits **44**(1), 32–48 (2009). https://doi.org/10.1109/JSSC.2008.2007145
13. Dixit, H.D., Pendharkar, S., Beadon, M., Mason, C., Chakravarthy, T., Muthiah, B., Sankar, S.: Silent data corruptions at scale. CoRR abs/2102.11245 (2021). https://arxiv.org/abs/2102.11245

14. Ejlali, A., Al-Hashimi, B.M., Eles, P.: A standby-sparing technique with low energy-overhead for fault-tolerant hard real-time systems. In: Proceedings of the 7th IEEE/ACM International Conference on Hardware/Software Codesign and System Synthesis, CODES+ISSS '09, pp. 193–202. Association for Computing Machinery, New York (2009). https://doi.org/10.1145/1629435.1629463
15. Eles, P., Izosimov, V., Pop, P., Peng, Z.: Synthesis of fault-tolerant embedded systems. In: 2008 Design, Automation and Test in Europe, pp. 1117–1122 (2008). https://doi.org/10.1109/DATE.2008.4484825
16. Eles, P., Izosimov, V., Pop, P., Peng, Z.: Synthesis of fault-tolerant embedded systems. In: 2008 Design, Automation and Test in Europe, pp. 1117–1122 (2008). https://doi.org/10.1109/DATE.2008.4484825
17. Ernst, D., Kim, N.S., Das, S., Pant, S., Rao, R., Pham, T., Ziesler, C., Blaauw, D., Austin, T., Flautner, K., Mudge, T.: Razor: a low-power pipeline based on circuit-level timing speculation. In: Proceedings. 36th Annual IEEE/ACM International Symposium on Microarchitecture, 2003. MICRO-36, pp. 7–18 (2003). https://doi.org/10.1109/MICRO.2003.1253179
18. Floyd, M.S., Drake, A.J., Berry, R.W., Chase, H., Willaman, R., Pena, J.: Voltage droop reduction using throttling controlled by timing margin feedback. In: 2012 Symposium on VLSI Circuits (VLSIC), pp. 96–97 (2012). https://doi.org/10.1109/VLSIC.2012.6243807
19. Gupta, M.S., Rangan, K.K., Smith, M.D., Wei, G.Y., Brooks, D.: DeCoR: a delayed commit and rollback mechanism for handling inductive noise in processors. In: 2008 IEEE 14th International Symposium on High Performance Computer Architecture, pp. 381–392 (2008). https://doi.org/10.1109/HPCA.2008.4658654
20. Gupta, S., Feng, S., Ansari, A., Blome, J., Mahlke, S.: StageNetSlice: a reconfigurable microarchitecture building block for resilient CMP systems. In: Proceedings of the 2008 International Conference on Compilers, Architectures and Synthesis for Embedded Systems, CASES '08, p. 1–10. Association for Computing Machinery, New York (2008). https://doi.org/10.1145/1450095.1450099
21. Henkel, J., Bauer, L., Becker, J., Bringmann, O., Brinkschulte, U., Chakraborty, S., Engel, M., Ernst, R., Härtig, H., Hedrich, L., Herkersdorf, A., Kapitza, R., Lohmann, D., Marwedel, P., Platzner, M., Rosenstiel, W., Schlichtmann, U., Spinczyk, O., Tahoori, M., Teich, J., When, N., Wunderlich, H.J.: Design and architectures for dependable embedded systems. In: 2011 Proceedings of the Ninth IEEE/ACM/IFIP International Conference on Hardware/Software Codesign and System Synthesis (CODES+ISSS), pp. 69–78 (2011). https://doi.org/10.1145/2039370.2039384
22. Henkel, J., Bauer, L., Dutt, N., Gupta, P., Nassif, S., Shafique, M., Tahoori, M., Wehn, N.: Reliable on-chip systems in the nano-era: lessons learnt and future trends. In: 2013 50th ACM/EDAC/IEEE Design Automation Conference (DAC), pp. 1–10 (2013). https://doi.org/10.1145/2463209.2488857
23. Henkel, J., Bauer, L., Zhang, H., Rehman, S., Shafique, M.: Multi-layer dependability: From microarchitecture to application level. In: 2014 51st ACM/EDAC/IEEE Design Automation Conference (DAC), pp. 1–6 (2014). https://doi.org/10.1145/2593069.2596683
24. Hochschild, P.H., Turner, P., Mogul, J.C., Govindaraju, R., Ranganathan, P., Culler, D.E., Vahdat, A.: Cores that don't count. In: Proceedings of the Workshop on Hot Topics in Operating Systems, HotOS '21, pp. 9–16. Association for Computing Machinery, New York (2021). https://doi.org/10.1145/3458336.3465297
25. Huang, L., Yuan, F., Xu, Q.: Lifetime reliability-aware task allocation and scheduling for MPSoC platforms. In: 2009 Design, Automation & Test in Europe Conference & Exhibition, pp. 51–56 (2009). https://doi.org/10.1109/DATE.2009.5090632
26. Huard, V., Cacho, F., Giner, F., Saliva, M., Benhassain, A., Patel, D., Torres, N., Naudet, S., Jain, A., Parthasarathy, C.: Adaptive wearout management with in-situ aging monitors. In: 2014 IEEE International Reliability Physics Symposium, pp. 6B.4.1–6B.4.11 (2014). https://doi.org/10.1109/IRPS.2014.6861106

27. Hyman, R., Bhattacharya, K., Ranganathan, N.: A strategy for soft error reduction in multi core designs. In: 2009 IEEE International Symposium on Circuits and Systems, pp. 2217–2220 (2009). https://doi.org/10.1109/ISCAS.2009.5118238
28. Kongetira, P., Aingaran, K., Olukotun, K.: Niagara: a 32-way multithreaded SPARC processor. IEEE Micro **25**(2), 21–29 (2005). https://doi.org/10.1109/MM.2005.35
29. Lee, H., Shin, H., Min, S.L.: Worst case timing requirement of real-time tasks with time redundancy. In: Proceedings Sixth International Conference on Real-Time Computing Systems and Applications. RTCSA'99 (Cat. No.PR00306), pp. 410–414 (1999). https://doi.org/10.1109/RTCSA.1999.811290
30. Lee, K., Shrivastava, A., Kim, M., Dutt, N., Venkatasubramanian, N.: Mitigating the impact of hardware defects on multimedia applications: a cross-layer approach. In: Proceedings of the 16th ACM International Conference on Multimedia, MM '08, pp. 319–328. Association for Computing Machinery (2008). https://doi.org/10.1145/1459359.1459402
31. May, M., Alles, M., Wehn, N.: A case study in reliability-aware design: a resilient LDPC code decoder. In: 2008 Design, Automation and Test in Europe, pp. 456–461 (2008). https://doi.org/10.1109/DATE.2008.4484723
32. Melhem, R., Mosse, D., Elnozahy, E.: The interplay of power management and fault recovery in real-time systems. IEEE Trans. Comput. **53**(2), 217–231 (2004). https://doi.org/10.1109/TC.2004.1261830
33. Mitchell, J., Henderson, D., Ahrens, G.: IBM POWER 5 processor-based servers: a highly available design for business-critical applications. Tech. rep., IBM white paper (2005). http://www-07.ibm.com/systems/includes/pdf/power5_ras.pdf
34. Pop, P., Izosimov, V., Eles, P., Peng, Z.: Design optimization of time-and cost-constrained fault-tolerant embedded systems with checkpointing and replication. IEEE Trans. Very Large Scale Integration Syst. **17**(3), 389–402 (2009). https://doi.org/10.1109/TVLSI.2008.2003166
35. Pyo, S.S., Lee, C.H., Kim, G.H., Choi, K.M., Jun, Y.H., Kong, B.S.: 45nm low-power embedded pseudo-SRAM with ECC-based auto-adjusted self-refresh scheme. In: 2009 IEEE International Symposium on Circuits and Systems, pp. 2517–2520 (2009). https://doi.org/10.1109/ISCAS.2009.5118313
36. Ranjbar, B., Hosseinghorban, A., Salehi, M., Ejlali, A., Kumar, A.: Toward the design of fault-tolerance-aware and peak-power-aware multicore mixed-criticality systems. IEEE Trans. Comput.-Aided Design Integrated Circuits Syst. **41**(5), 1509–1522 (2022). https://doi.org/10.1109/TCAD.2021.3082495
37. Sabry, M.M., Atienza, D., Catthoor, F.: OCEAN: An optimized HW/SW reliability mitigation approach for scratchpad memories in real-time SoCs. ACM Trans. Embed. Comput. Syst. **13**(4s) (2014). https://doi.org/10.1145/2584667
38. Salehi, M., Khavari Tavana, M., Rehman, S., Shafique, M., Ejlali, A., Henkel, J.: Two-state checkpointing for energy-efficient fault tolerance in hard real-time systems. IEEE Trans. Very Large Scale Integration Syst. **24**(7), 2426–2437 (2016). https://doi.org/10.1109/TVLSI.2015.2512839
39. Vayrynen, M., Singh, V., Larsson, E.: Fault-tolerant average execution time optimization for general-purpose multi-processor system-on-chips. In: 2009 Design, Automation & Test in Europe Conference & Exhibition, pp. 484–489 (2009). https://doi.org/10.1109/DATE.2009.5090713
40. Vayrynen, M., Singh, V., Larsson, E.: Fault-tolerant average execution time optimization for general-purpose multi-processor system-on-chips. In: 2009 Design, Automation & Test in Europe Conference & Exhibition, pp. 484–489 (2009). https://doi.org/10.1109/DATE.2009.5090713
41. Zhu, D., Aydin, H.: Reliability-aware energy management for periodic real-time tasks. IEEE Trans. Comput. **58**(10), 1382–1397 (2009). https://doi.org/10.1109/TC.2009.56

Chapter 8
Cross-Layer Reliability and Variability Mitigation Framework: Experiments

Abstract To evaluate the effectiveness, constraints, and costs of the cross-layer reliability and variability mitigation framework, a holistic simulation flow needs to be developed, in which the simulations of each layer are incorporated. Here, we developed a workload-dependent simulation framework for each level of mitigation. Combining individual simulations, we can perform cross-layer simulations to reveal the interdependency among layers, as well as the figures of merits in the reliability of the holistic system. The simulation framework can be further used for co-simulation and co-optimization of system parameters across all layers for maximizing reliability while minimizing cost.

This chapter is structured as follows: The overall simulation framework of the cross-layer reliability and variability mitigation framework is introduced in Sect. 8.1. The following sections show the experiments on the system reliability with different layers of the framework. In Sect. 8.2, we show a workload-dependent error simulation flow to evaluate the effect of the circuit-layer mitigation. Then, Sect. 8.3 demonstrates a cross-layer co-simulation combining the register layer and the processor layer to reveal their interdependencies. Finally, the effect of optimal checkpointing on the system is analyzed in Sect. 8.4, showing how changing system parameters can improve the reliability of the system.

8.1 Cross-Layer Simulation Framework

To simulate the effect of the entire cross-layer reliability and variability mitigation framework, multiple simulation methods for different purposes in different layers need to be integrated. Figure 8.1 shows the overall simulation framework required for the experiments on the whole framework. In our work, the essential parts of the simulation framework are implemented, while the rest of the framework and the integration of the holistic framework are yet to be developed. Nevertheless, the principles of establishing the cross-layer simulation framework are demonstrated, and important properties of cross-layer mitigation approaches are revealed through the simulation, which we will demonstrate in the following section.

J.-Y. Lin et al., *Workload Dependent Mitigation Approaches for Performance Variability*, https://doi.org/10.1007/978-3-031-85624-2_8

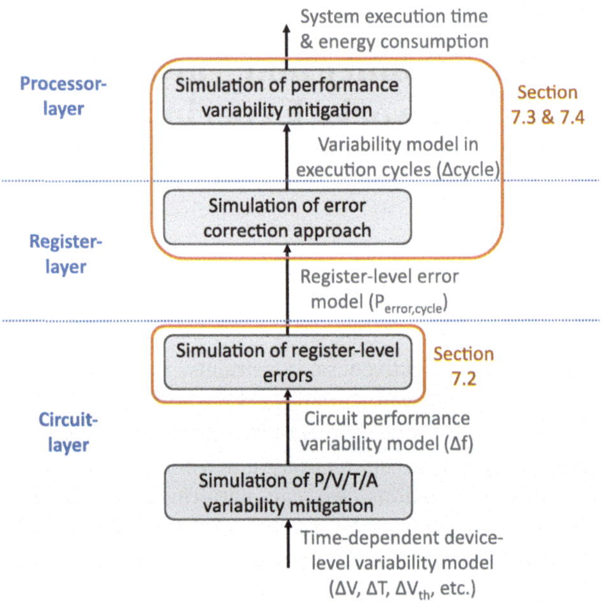

Fig. 8.1 Simulation framework of the cross-layer reliability and variability mitigation

At the beginning of the simulation framework, device-level variability models are required to generate the time-zero and time-dependent variability in process, voltage, temperature, and aging. The variability models vary with the addressed variability sources, expressed in their own physical metrics (ΔV, ΔT, ΔV_{th}, etc.). They can be either statistical models such as the V_{th} distribution of the process variation [1] or time-dependent traces of variability metrics such as BTI-induced V_{th} fluctuation over time [2].

Then, the simulation of the device-level variability mitigation is required to simulate the resulting circuit performance variability under the applied device-level variability. The simulation approach depends on the types of device-level variability and the applied mitigation approaches. In general, most of the device-level variability and mitigation approaches result in the time-dependent clock frequency change, so the circuit performance variability can be modeled as frequency shift (Δf) over time. For time-zero variability, variation-aware static timing analysis has been developed to accurately simulate the performance variability of logic gates and digital circuits [3, 6]. However, there is a lacking of mature approaches to accurately simulate the influence on the circuits from the time-dependent variability such as aging effect. Therefore, research has been conducted in order to fill this gap [2, 5], and we look forward to the development of an accurate time-dependent variability simulation flow.

After the performance variability of the circuits is known, the simulation of register-level errors is used to obtain the occurrence rate of the errors in the registers, which is modeled by the error probability per cycle ($P_{\text{error,cycle}}$). To serve this purpose, in Sect. 8.2 we will demonstrate a workload-dependent simulation flow which can simulate the register errors in a circuit when running a certain workload. In the simulation, a constant clock frequency shift (Δf) is injected into the circuit, causing register errors which can be read out as the results.

Once we know the amount of register-level errors in the circuit, the simulation of error correction approaches is used to simulate the error correction events in the system, which we will demonstrate in Sect. 8.3. In the simulation, the register-level errors are generated based on a constant error probability ($P_{\text{error,cycle}}$) for each cycle and injected into the workload to be corrected by the error correction approach. As a result, the execution time overhead caused by the error correction is obtained and modeled as the variability in the execution cycles (Δcycle). In theory, the execution cycle overheads of correcting each segment of the workload follow the Geometric distribution which we have derived in Sect. 7.3.3.

Afterward, the variability in execution cycles (Δcycle) is injected into the workload used as the input of the simulation of performance variability mitigation. The simulation flow has been described in Sect. 6.1. Finally, the results of this simulation show the execution time and the energy consumption of the system running the workload if all the addressed variability and errors are properly handled.

To fully simulate the interdependencies across layers, the individual simulation flows described above need to be integrated, in which the output variability and models of the former layer should be applied as the input of the latter layer. In the simulation framework, we have accomplished the integration of the simulations of the error correction approach and the performance variability mitigation, in which the exact execution cycle overhead of each workload segment from error correction is applied directly to the inputs of the performance variability mitigation. The integration of both simulation flows allows us to experiment with the interdependencies of the two mitigation approaches, which we will elaborate on in Sects. 8.3 and 8.4.

8.2 Simulation of Register-Level Errors

To analyze the reliability of systems, it is crucial to evaluate the number of register-level errors in the system under different variability and operating conditions, either in design time or in run-time. Therefore, we developed a workload-dependent error simulation flow to acquire the error occurrences during the running of a certain workload. The overall simulation flow is shown in Fig. 8.2.

At the beginning, variability-aware timing analysis should be performed to how much the gate delays and the clock frequency of the circuit are shifted by the time-zero/time-dependent device-level variability. As mentioned in Sect. 8.1, the time-zero/time-dependent variability models depend on the addressed variability sources and are yet to be integrated into our simulation flow. Instead, in this simulation a constant clock frequency shift is assumed to be the impact of the variability on the circuit. While the source of the frequency shift is not specified, it could originate

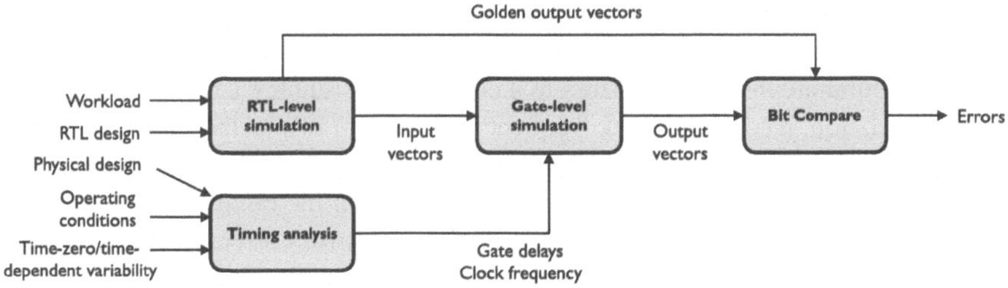

Fig. 8.2 Workload-dependent error simulation flow

from device-level variability with a timescale longer than the workload, such as process variation, voltage shift, temperature change, or long-term aging effect. It could also occur as a result of the modulation by the mitigation approaches of the device-level variability, such as adaptive clocking for voltage droop prevention.

Meanwhile, RTL-level simulation is performed on the RTL design of the target circuit running the target workload, where the golden input and output vectors of all registers in the design are extracted. Then, gate-level simulation is performed, based on the gate delays and the clock frequency after the effect of the variability. In the simulation, the gate delay propagation is captured in ps-level precision, so timing violations and register-level errors are generated and caught in the output vectors. Finally, by comparing the gate-level output vectors and golden output vectors, the number and the locations of register-level errors are acquired.

The error simulation flow enables error rate analysis under different variability and operating conditions. As an example, we analyzed the impact of the design margin on the error rate of the ALU pipeline stage of the Ariane core. In this experiment, we swept across different clock frequencies to represent the scenario in which the clock frequency is changed. No other variations in voltage, temperature, or aging are applied. The RTL-level model used in the simulation is the same as in Sect. 6.1, and the gate-level model and physical implementation are implemented in commercial 28 nm technology. The applied workload is the same lower-band quantization of ADPCM encoding workload as in Sect. 6.2, in which the length is about 3.6 million cycles.

In our analysis, we define the cycle-based error rate ($P_{\text{error,cycle}}$) as the number of erroneous cycles to the total number of cycles. A cycle is erroneous if any bit in the output vectors has the wrong value in this cycle. Figure 8.3 shows the simulation results. The worst-case frequency calculated by static timing analysis is 2.61 GHz, defined by the critical path delay. However, if the critical path is not activated during the entire workload, the circuit can run at a higher clock frequency without any errors. In our case, the clock frequency can be increased without observing any errors until 2.73 GHz, which is 4.6% higher than the worst-case frequency. Then, if the clock frequency is further increased, the error rate is increased rapidly. At 2.78 GHz, a cycle-based error rate is already larger than 1%. In summary, the error rate is shown to be highly sensitive to clock frequency, which modulates the design margin of the circuit. The high sensitivity of error rate to operating conditions also

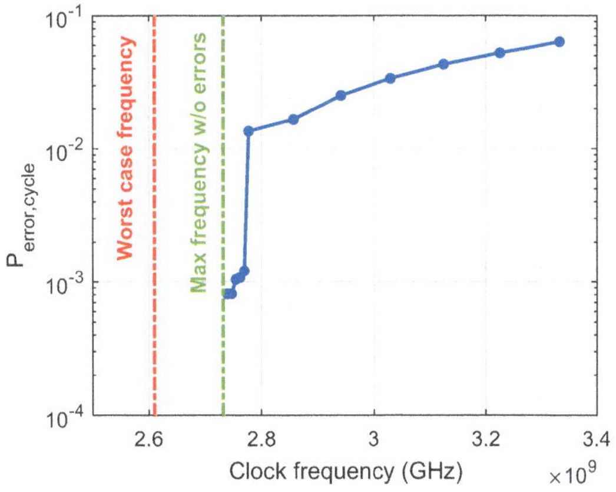

Fig. 8.3 Error rate per cycle under different clock frequencies

shows the importance of accurate error rate evaluation, since any tiny discrepancy in operating conditions could result in significant degradation of reliability.

8.3 Cross-Layer Interdependency Analysis

To evaluate the overall effect when multiple mitigation techniques work together, it is essential to understand the interdependencies across layers and different mitigation approaches. As an example, we study and simulate the effect of combining CRR in the register layer and performance variability mitigation in the processor layer. Specifically, Fig. 8.4 shows how the two mitigation approaches can collaborate to meet the dual reliability requirements of functional correctness and timing guarantees. When CRR is applied to correct the errors in the workload, the execution time overhead can result in deadline misses. Therefore, performance variability mitigation is applied to increase processor speed to offset the execution time overhead, so that timing guarantees can be ensured.

The interdependency of the two approaches resides in the execution time overhead of CRR. When the error rate is high, the execution time overhead will be large, so the processor speed needs to be increased more dramatically to compensate for the overhead, and thus the energy consumption will be large. Furthermore, if the error rate is too high, the workload can be flooded with rollback events, so that even the highest processor speed cannot offset the time overhead. In this scenario, timing guarantees cannot be ensured anymore and deadline misses can occur. In summary, Fig. 8.5 shows the outcomes of the collaboration of the two approaches, in which the energy is increased with the error rate, and the deadline hit rate is decreased with the error rate. In the following subsections, we will show how the interdependences can be simulated.

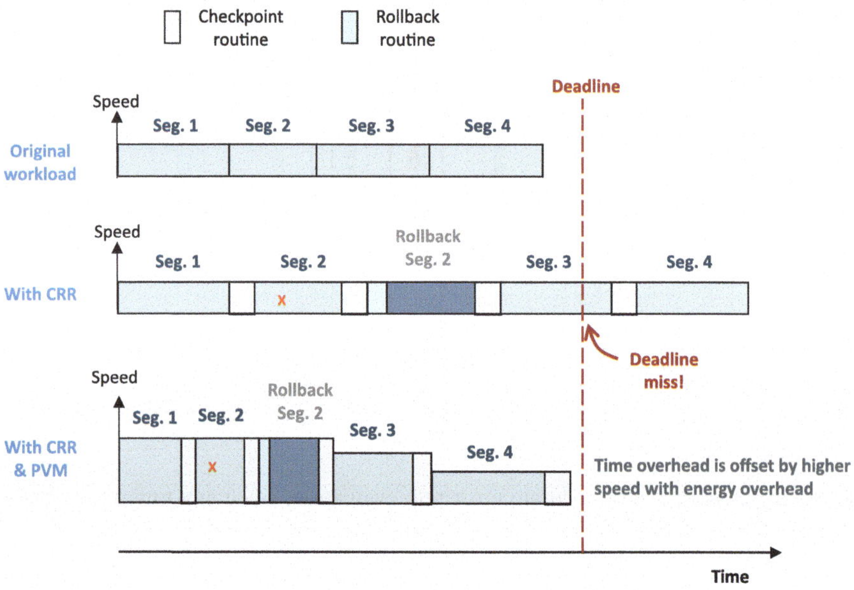

Fig. 8.4 Illustration of an example of how CRR and performance variability mitigation can collaborate to meet the dual reliability requirements

Fig. 8.5 Interdependency of CRR and performance variability mitigation

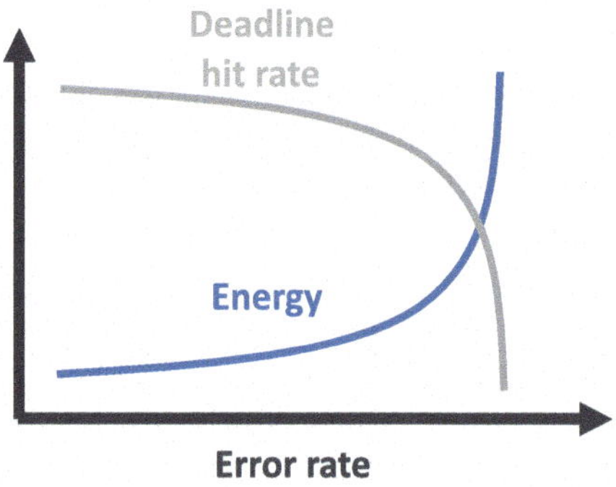

8.3.1 *Simulation Flow and Experimental Setup*

To analyze the effectiveness and cost of the holistic system in meeting the dual reliability requirements simultaneously, it is necessary to establish a cross-layer simulation framework which integrates the simulation of every individual technique. Therefore, we developed a simulation framework which combines the simulation of the circuit-layer CRR approach and the processor-layer performance variability mitigation approach.

Figure 8.6 shows the overall flow of the simulation framework. First, the target workload is profiled with RTL-level simulation to the original execution time

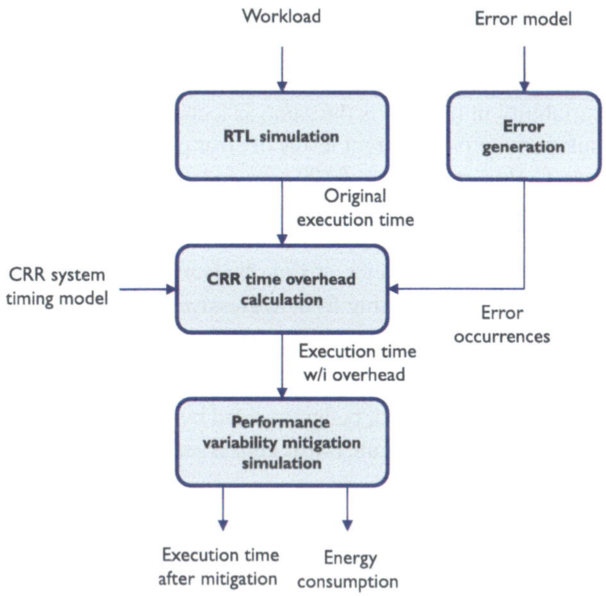

Fig. 8.6 Cross-layer simulation flow incorporating the register-layer CRR and the processor-layer performance variability mitigation

without errors. Then, errors are generated inside the workload based on the error model. Here, we assume a static error rate model in which the error probability is the same for each cycle over the entire workload, while the error simulation flow in Sect. 8.2 can also be integrated for more sophisticated simulation.

Afterward, to simulate the effect of the CRR approach, a system timing model is created by abstracting the OCEAN approach described in Sect. 7.3.2 [4]. In this model, the checkpoint routine and the rollback routine are assumed to be 100 cycles and 48 cycles, respectively, regardless of the computation phases. And the recomputation is assumed to take the same execution time as the original computation phase. The number of rollbacks is randomly generated for each computation phase with the probability distribution of Eq. 7.2, based on the error probability specified by the error rate model. By summing up the execution time of the checkpoint routines, the rollback routines, and the recomputation, the time overhead of each computation phase of the workload can be calculated.

Next, the execution time after CRR is used as the input of the simulation of performance variability mitigation. For this purpose, we use the workload-dependent multi-timescale mitigation approach in Chap. 5 and the cycle-accurate simulation flow in Chap. 6. The simulation flow generates the execution time and energy consumption after the mitigation as the final outcomes of the whole framework. In the calculation of the energy consumption, the energy of the checkpoint and rollback routines are considered to consume the same energy per cycle as the TN they belong to.

In the experiment, we use the same Ariane core processor and ADPCM lower-band quantization workload as the experiment in Chap. 6. The segmentation of workload is the same for both mitigation approaches. In other words, a TN in performance variability mitigation is the same as a computation phase in CRR. We performed simulations across different levels of error probability. For each level, we performed Monte Carlo simulations of 100 runs and calculated the average results.

In the scheduling algorithm, timing budgets can be deliberately enlarged to absorb the CRR time overhead, so that deadline misses can be prevented. To compare different algorithms and policies of time budget assignment, four algorithms are selected in our simulation, ranging from aggressive ones to conservative ones.

- **DS**: dynamic scenario-based ***(most aggressive)***
- **DS 1.5×**: dynamic scenario-based, budgets scaled by 1.5×
- **DS 2×**: dynamic scenario-based, budgets scaled by 2×
- **WCET**: worst-case execution time ***(most conservative)***

The DS-based approach is the most aggressive algorithm, which can derive a tight budget with full guarantees under the error-free condition at run-time. On the other hand, worst-case execution time is the most conservative algorithm. Moreover, two additional variants of the DS-based algorithms are created, in which the budgets for all TNs are enlarged by 1.5 times and 2 times, respectively. These algorithms aim to reflect the influence of the aggressiveness of algorithms on reliability and costs, which has been discussed in Sect. 7.4.

8.3.2 Results and Discussion

Figure 8.7 shows the number of rollbacks increases with the error probability in our demonstrated system. The rollbacks result in increased numbers of executed cycles, shown in Fig. 8.8. The rollbacks and the execution cycles are only influenced by the CRR systems, regardless of the algorithm used in performance variability mitigation. The extra cycles apart from the original workload are approximately linear to the number of rollbacks. The trend of the rollbacks and the total cycles is consistent with the execution time overhead of Figs. 7.8 and 7.9, which proves the correctness of the theoretical calculation in Sect. 7.3.3.

The error rate wall of this system can be derived with the execution cycles and the highest utilization over the original workload, as described in Sect. 7.5. In our ADPCM lower-band quantization workload, the highest utilization is calculated based on the frame with the highest processing time over the entire workload, which results in a utilization of 0.54. With Eq. 7.7 and Fig. 8.8, the error rate wall is estimated to be at the error rate of 4.5×10^{-6} each cycle.

Figure 8.9 shows the deadline hit rate under different error probability conditions in our demonstrated system. It shows that the error rate wall is approximately located at the boundary of whether the 100% hit rate can be ensured with conservative algorithms (DS 2.0×, WCET). Before the error rate wall, the execution

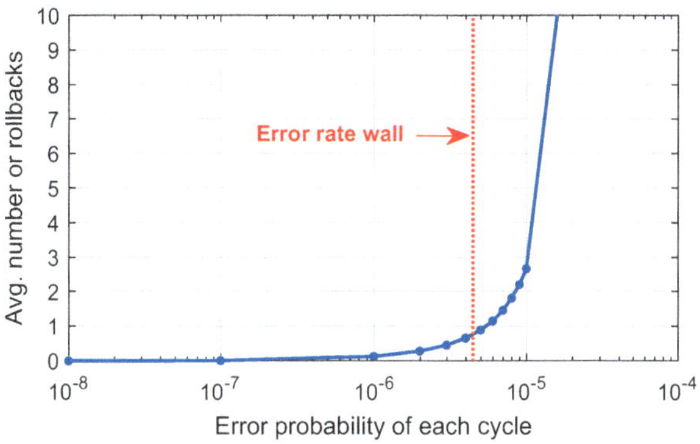

Fig. 8.7 Average rollbacks of every computation phase under different error probability conditions

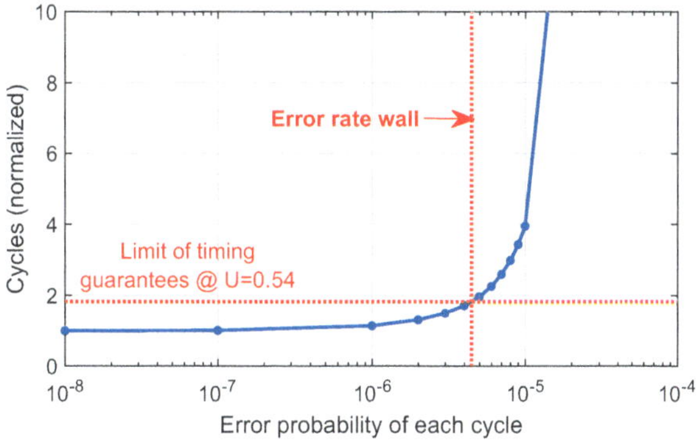

Fig. 8.8 Total executed cycles normalized to the cycles of the original workload, under different error probability conditions

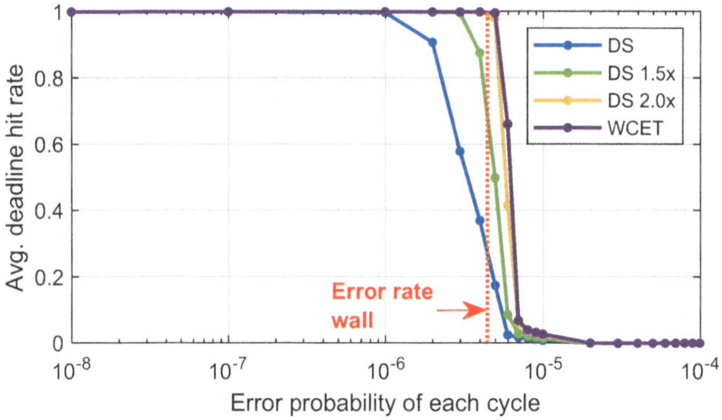

Fig. 8.9 Deadline hit rate under different error probability conditions

time overhead could be offset by higher processor speed, so deadline misses could still be completely prevented. However, beyond the error rate wall, the rollbacks are so frequent that even the highest processor speed cannot save the deadlines from being missed. Therefore the hit rate rapidly drops to almost zero in a small window of error probability, which shows the high sensitivity of timing guarantees to error

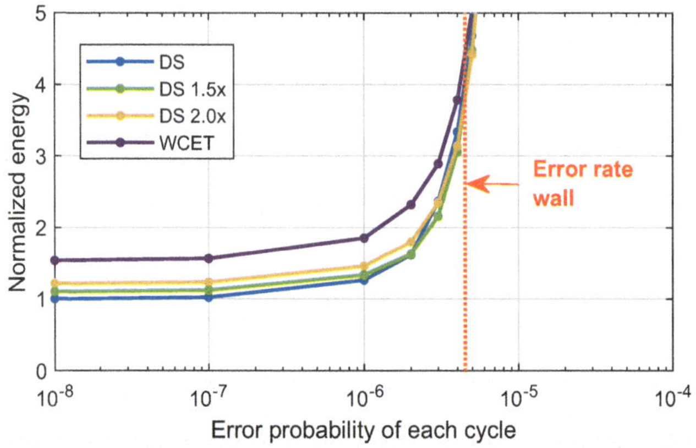

Fig. 8.10 Total energy consumption normalized to the energy consumption of the original workload, under different error probability conditions

probability. If the error probability is even higher, the number of rollbacks could increase to more than ten rollbacks per computation phase, which means that the workload would be flooded with rollbacks. Dealing with such a huge number of rollbacks is impractical for a reasonable system. These results show that the error rate wall defines the inherent system limit of the conditions that both reliability requirements can be fulfilled, which cannot be improved by the algorithms.

A comparison among scheduling algorithms shows that conservative algorithms bring higher deadline hit rates. Aggressive algorithms (DS, DS 1.5×) can have deadline misses even before the error rate wall is reached, due to the insufficient time slack left for absorbing the execution time overhead. However, beyond the error rate wall, all deadline hit rates converge to zero regardless of the algorithms. These results also prove that the error rate wall is the inherent system limit independent of the algorithms.

Figure 8.10 shows the energy consumption under different error probability conditions. Near the error rate wall, the energy consumption increases rapidly so that it is about four times the original energy consumption at the error rate wall. The sources of the energy overhead are twofold. First, running the checkpoints, rollbacks, and recomputations, all result in extra energy. Second, the higher processor speed for offsetting time overhead leads to higher power dissipation and thus consumes more energy. As for the algorithms, conservative algorithms have higher energy overheads when the error probability is low. However, when the error probability is high, aggressive algorithms may have deadline misses, which make the processor speed up to recover from the negative slack. Therefore, the advantage of aggressive algorithms in using slower processor speed diminishes, so the energy consumption of different algorithms becomes similar.

The position of the error rate wall is strongly dependent on system parameters, such as the processor speed, the granularity of checkpointing, etc. To know whether a system can ensure its reliability requirement, one can check whether the position of the error rate wall is further than the error probability, which can be simulated by the flow we showed in Sect. 8.2. Since the system reliability depends on both the error rate wall and the error probability, the ways of improving system reliability are involved with affecting these two factors. One way is to reduce the error probability through technology improvement (e.g., advanced lithography) or circuit design (e.g., ECC or DMR). However, as discussed in Sect. 7.3.1, these error-tolerant circuit design techniques can lead to significant area, performance, and energy overheads. The other way is to improve the tolerance of errors (i.e., moving the error rate wall further) by optimizing the system. For example, minimizing execution time overhead by optimizing the number of checkpoints can improve reliability, as we will show in the next section. Co-simulation and co-optimization of all system parameters across all layers and techniques are required in order to maximize the reliability and minimize the cost of the system.

8.4 Analysis with Optimal Checkpointing

In Sect. 7.3.4, we have shown that to minimize execution time overhead, the optimal numbers of checkpoints are different for different error probability conditions. With the minimum execution time overhead, both the deadline misses and the energy costs are reduced in performance variability mitigation, so reliability can be improved. Here, we will analyze how the error rate wall can be moved by selecting different numbers of checkpoints, which is an example of how system parameters affect reliability.

8.4.1 Experimental Setup

We used the same simulation flow and experiment setup as in Sect. 8.3. However, in this experiment, a TN is segmented into multiple computation phases. In the calculation of execution time and energy consumption, computation phases in the same TN are assumed to be homogeneous. For example, if a TN is segmented into ten computation phases, each phase takes 10% cycles and 10% energy consumption of the original TN under the same processor speed. As in the original CRR mechanism, a 100-cycle checkpoint routine is added at the end of each computation phase. If any error occurs, a 48-cycle rollback routine is added, and recomputation is performed. By Eq. 7.5, we calculated the optimal number of checkpoints per TN based on the average length of all TNs. The relation of the error probability and the optimal number of checkpoints are shown in Fig. 8.11. Besides the case of the optimal number of checkpoints, we also simulated the case of 1, 10, 100, 1000, and 2000 checkpoints per TN for comparison. The DS $2\times$ algorithm of the last

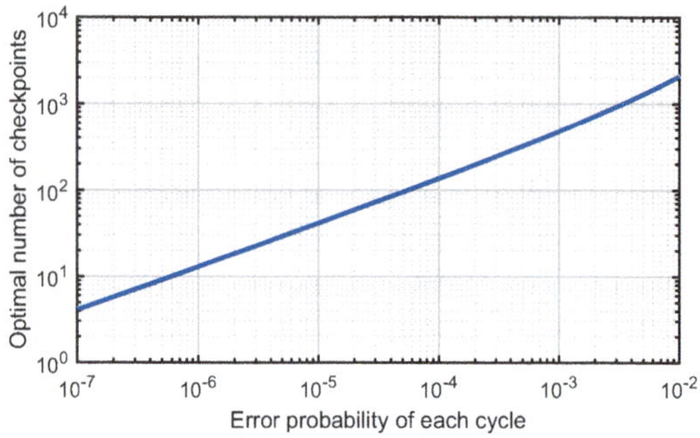

Fig. 8.11 Optimal number of checkpoints per TN under different error probability conditions

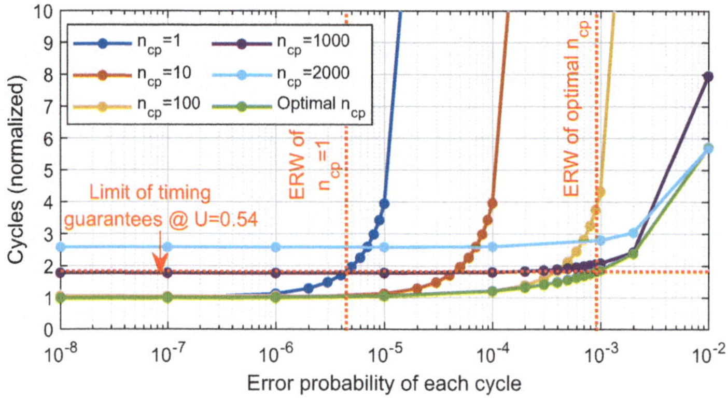

Fig. 8.12 Total executed cycles with different numbers of checkpoints per TN (n^{cp}), under different error probability conditions. The numbers of executed cycles are normalized to the cycles of the original workload. The red dash line indicates the error rate wall under the optimal n^{cp} condition

section is used in this experiment, which is a relatively conservative algorithm that can approach the system limit in mitigating the execution time overhead.

8.4.2 *Results and Discussion*

In the simulation results, Fig. 8.12 shows that the numbers of total executed cycles are highly dependent on the number of checkpoints. When the number of checkpoints increases from 1 to 100, the overhead of cycles is remarkably reduced at the high error probability over 10^{-6} per cycle, since the recomputed workload segment required for each error occurrence is largely reduced by 100 times. As a result, the error tolerance is approximately improved by two orders.

However, when the number of checkpoints further increases to 1000 and 2000, the cycles of checkpoints become significant and add up to 80% and 160% cycle overhead, respectively, to the original workload, significantly increasing the total

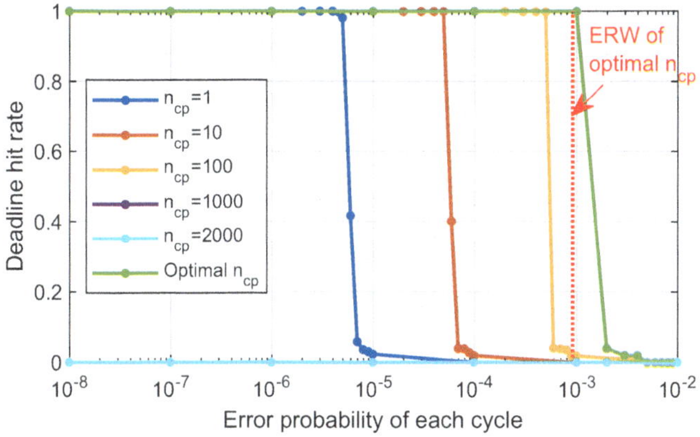

Fig. 8.13 Deadline hit rate with different numbers of checkpoints per TN (n^{cp}), under different error probability conditions. The curves of $n^{cp} = 1000$ and optimal n^{cp} overlap with each other, yielding similar deadline hit rates

execution cycles even at low error probability. Such a high overhead would greatly degrade the performance and energy efficiency of a system. Even though these schemes have smaller overheads at the high error probability over 10^{-3} per cycle, the overhead is still too high for a feasible system. Furthermore, in the 2000-checkpoint scheme, the high overhead makes the number of cycles exceed the limit of ensuring timing guarantees in all error probability ranges. Thus, the 2000-checkpoint scheme is not feasible for system reliability. Also, any schemes of more than 2000 checkpoints only increase the overhead even more, so they cannot meet the system reliability requirements.

Nevertheless, if the optimal number of checkpoints is applied for each error probability condition, the cycle overhead can be minimized. Also, the trend of the number of cycles does not rise as quickly with the error probability as in the cases of the fixed number of checkpoints. The smaller sensitivity to the error probability means that the system is more robust to miscalculation and variation of the error rate, which also benefits the overall reliability of the system.

The minimized cycle overhead leads to better deadline hit rates as shown in Fig. 8.13. With the increase in the number of checkpoints from 1 to 1000, the error rate wall is pushed forward to 9×10^{-4} error probability per cycle, about 200 times to the original ERW of $n^{cp} = 1$. The results are consistent with the observation of Fig. 8.12 that the 1000-checkpoint scheme has the smallest overhead among all fixed-checkpoint schemes at around the 10^{-3} error probability condition. However, when the number of checkpoints further increases to 2000, the deadline hit rate drops to 0 in all error probability conditions, since the overhead from the checkpointing routines is too large to be mitigated by the highest processor speed even without any errors. The failure to meet deadlines in the 2000-checkpoint scheme is in agreement with the observation in Fig. 8.12 that the execution cycles exceed the limit. Furthermore, we suppose that any schemes above

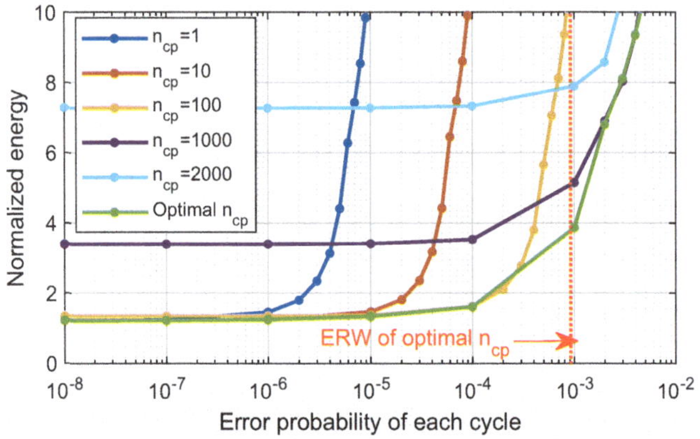

Fig. 8.14 Total energy consumption with different numbers of checkpoints per TN (n^{cp}), under different error probability conditions. The energy consumptions are normalized to the one of the original workload

2000 checkpoints only increase the overhead, so the deadline hit rate will still stay at 0.

With the optimal number of checkpoints, the deadline hit rates can always be maximized in any error probability condition, and the error probability the system can operate without deadline misses is the highest. While the case of 1000 checkpoints also yields similar deadline hit rates, the overhead of checkpointing routines increases the execution cycles too much under the low error probability condition, consuming significant energy which is discussed below.

The energy consumptions in Fig. 8.14 show similar effects. Increasing the number of checkpoints to 100 reduces energy overhead at high error probability conditions, so the system can operate at 100× higher error probability without energy rising sharply. However, with 1000 checkpoints per TN, the large overhead of checkpoints increases energy consumption to more than three times the original energy consumption, which greatly degrades the energy efficiency of the systems. For the 2000-checkpoint scheme, the energy consumption is even more than seven times, and further increasing the checkpoints only enlarges the energy consumption. With optimal checkpointing, energy consumption can always be minimized at any error probability condition. Also, the sensitivity of energy consumption to the error probability is smaller.

In conclusion, we demonstrated how optimal checkpointing can remarkably improve reliability and reduce system costs in execution time and energy consumption. Therefore, checkpointing needs to be adaptive to the error rate in run-time. To realize this, it is crucial to develop well-supported software and firmware structures for adaptive checkpointing and accurate run-time error rate estimation in hardware.

8.5 Summary

We developed a workload-dependent simulation framework which combines individual simulations of different layers to reveal the holistic reliability and costs of the system. First, we demonstrated how a workload-dependent error simulation flow can evaluate the error rate of the registers under different variability and operation conditions. Then, we showed a cross-layer simulation flow that incorporates register-level CRR and processor-level performance variability mitigation. The flow enables analysis of system reliability in meeting both functional correctness and timing guarantees, as well as the underlying energy overheads in meeting these requirements. The analysis proves the concept of the "error rate wall," the error rate limit in which the system can ensure the dual reliability requirements. Finally, we showed the error rate wall can be improved by using the optimal number of checkpoints, which is an example of how optimizing system parameters can improve the reliability of the entire system.

Our work can be further extended to a holistic framework incorporating all mitigation approaches from the device to application layers to ensure multiple reliability requirements. This framework could enable co-simulation and co-optimization of system parameters across all layers to maximize reliability and minimize costs. In this way, we can improve the designs of highly reliable systems by holistically considering all reliability requirements.

References

1. Kuhn, K.J., Giles, M.D., Becher, D., Kolar, P., Kornfeld, A., Kotlyar, R., Ma, S.T., Maheshwari, A., Mudanai, S.: Process technology variation. IEEE Trans. Electron Devices **58**(8), 2197–2208 (2011). https://doi.org/10.1109/TED.2011.2121913
2. Mishra, S., Weckx, P., Lin, J.Y., Kaczer, B., Linten, D., Spessot, A., Catthoor, F.: Fast & accurate methodology for aging incorporation in circuits using adaptive waveform splitting (AWS). In: 2020 IEEE International Reliability Physics Symposium (IRPS), pp. 1–5 (2020). https://doi.org/10.1109/IRPS45951.2020.9129351
3. Mutlu, A., Le, J., Molina, R., Celik, M.: A parametric approach for handling local variation effects in timing analysis. In: 2009 46th ACM/IEEE Design Automation Conference, pp. 126–129 (2009). https://doi.org/10.1145/1629911.1629949
4. Sabry, M.M., Atienza, D., Catthoor, F.: OCEAN: an optimized HW/SW reliability mitigation approach for scratchpad memories in real-time SoCs. ACM Trans. Embed. Comput. Syst. **13**(4s) (2014). https://doi.org/10.1145/2584667
5. van Santen, V.M., Amrouch, H., Henkel, J.: New worst-case timing for standard cells under aging effects. IEEE Trans. Device Mater. Reliab. **19**(1), 149–158 (2019). https://doi.org/10.1109/TDMR.2019.2893017
6. Walla, S.: Primetime® advanced OCV technology. Tech. rep., Synopsys white paper (2009). https://www.synopsys.com/content/dam/synopsys/implementation&signoff/white-papers/PrimeTime_AdvancedOCV_WP.pdf

Chapter 9
Conclusion and Future Work

Abstract In this chapter, we summarize the main messages and contributions of this book and discuss the future perspectives of workload-dependent mitigation approaches. The list of other publications of this research is also included.

9.1 Main Messages

The main focus of this book has been on the performance variability mitigation approaches to ensuring timing guarantees. On this topic, we conclude the general messages of this dissertation as follows:

- In Chap. 2, we discussed a wide range of variability sources from the device/circuit, architecture, and software levels. With the development of technology and architecture, variability becomes larger and more complex. Therefore, the guardbanding approach is no longer sufficient, and run-time mitigation approaches are applied to effectively tackle the variability.
- In Chap. 3, we discussed the principles and the components of performance variability mitigation approaches. The state-of-the-art approaches use the WCET-based algorithm and the DVFS knob. However, they are no longer sufficient for highly dynamic workloads, since WCET models are too simplified for modern complicated architectures and DVFS has long switching times.
- In Chap. 4, we present a realistic case study of an improved PID controller that mitigates performance variation with a DVFS response. The controller operates on finer granularity, at the thread node level. The basis of this new approach utilizes the system scenario methodology. In addition, a run-time adjustment on this methodology to adapt to performance variability norms is shown, creating an adaptive scenario scheme and achieving notable energy gains. Still, however, performance variation is managed with a reactive response, and no timing guarantees are yet delivered.
- The *workload-dependent multi-timescale mitigation approach* proposed in Chap. 5 can ensure timing guarantees with two features. The HDP knob pushes the time granularity to the sub-millisecond level, and the dynamic scenario-based

J.-Y. Lin et al., *Workload Dependent Mitigation Approaches for Performance Variability*, https://doi.org/10.1007/978-3-031-85624-2_9

algorithm enables run-time calculation of the tight upper bound of execution time. Then in Chap. 6, we showed that this approach can ensure full timing guarantees with 32% energy saving compared to the conventional WCET-based method with a single-DVFS knob.

- The *cross-layer variability mitigation framework* proposed in Chap. 7 can be a joint effort of individual mitigation approaches for different objectives in different layers, ensuring multiple reliability objectives in a cost-efficient way. In our framework, we use three mitigation approaches (circuit-level variability mitigation, register-level checkpointing and rollback-recovery, and processor-level performance variability mitigation) to simultaneously ensure functional correctness and timing guarantees.
- In Chap. 8 we showed that in a multi-objective reliability solution, interdependencies exist among different reliability objectives. In our cross-layer mitigation framework, the phenomenon of the *error rate wall* defines the maximum error rate which can be tolerated while ensuring timing guarantees. Furthermore, by optimizing system parameters (e.g., checkpointing granularity), it is possible to improve the error rate wall.

9.2 Contributions

In the following, we conclude the main contributions of our research:

- In Chap. 2, we gave a complete survey on all the variability sources of all hierarchy levels in modern systems of integrated circuits. These variability sources are classified from multiple aspects, including hierarchy levels (device and circuit, architecture, software), time span (time-zero, time-dependent), and influenced metrics (process, voltage, temperature, aging, performance variability). The general methods of tacking variability have also been discussed. This survey provides a broad picture of all kinds of variability problems that need to be tackled in modern systems and platform.
- In Chap. 3, we gave an overview of the start-of-the-art of mitigation approaches for performance variability. We summarized the common structure of run-time mitigation approaches and its necessary components. We also reviewed the mitigation approaches in literature and made a clear distinction between the approaches with and without timing guarantees. This overview can serve as the fundamentals which every mitigation approach should be developed upon.
- We have studied an enhanced version of the PID scheme that operates on the RTS level (Chap. 4). Given that modern applications typically consist of numerous RTSs, we have utilized the system scenario approach to group these RTSs into a more manageable set of scenario clusters. Through simulations, we have verified the ability of our enhanced scheme. The scenario clustering takes place offline, during the profiling of the application; a run-time, re-clustering policy through a novel, adaptive scenario approach is also presented. Simulation results comparing this scheme to the system scenarios and to a static frequency

guardband technology highlight the benefit of the former methodology both in terms of performance and energy response.

- In Chaps. 5 and 6, we have proposed a novel workload-dependent multi-timescale mitigation approach, which can mitigate performance variability with guarantees as required in all the three application classes discussed earlier: time-critical, time-sensitive, and general with latency bounds. It tackles architecture-level and software-level performance variability of different timescales while minimizing energy consumption. This approach has two salient features. First, a novel fast-switching system knob, HDP, is implemented which can push the time granularity of mitigation to the sub-millisecond level. Second, the scheduling algorithm is designed on the principle of dynamic scenarios, which can derive the tight upper bound of execution time to prevent deadline misses.
- In Chap. 6, we have developed a cycle-accurate simulation flow to evaluate the mitigation approach. Unlike the simplified processor models used in normal real-time scheduling research, this flow can capture the performance variability down to cycle-level granularity. Important processor metrics (execution time, energy consumption, etc.) can also be reported by the flow to evaluate the reliability and costs of different performance variability mitigation approaches.
- In Chaps. 7 and 8, a cross-layer reliability and variability mitigation framework is proposed to ensure both function correctness and timing guarantees at the same time. The framework consists of three mitigation approaches, including a circuit-level variability mitigation approach, register-level checkpointing and rollback-recovery, and processor-level performance variability mitigation.
- In Chap. 8, co-simulation of the layers in the aforementioned framework shows the holistic system impact and the system reliability limit in ensuring the dual reliability objectives. Also, by this cross-layer analysis, the interdependencies of different reliability objectives are revealed. Specifically, the system demonstrates the phenomenon of the "error rate wall," the error rate limit beyond which the functional correctness and timing guarantees cannot be ensured at the same time. We also show that changing system parameters improve the error rate wall. Therefore, the overall system reliability can be enhanced by optimizing system parameters.

9.3 List of Publications

The research contributions presented in this book have been published in the following work:

- Nikolaos Zompakis, M. Noltsis, D. Rodopoulos, F. Catthoor, and D. Soudris, "Energy Efficient Adaptive Approach for Dependable Performance in the presence of Timing Interference", in *Proceedings of the on Great Lakes Symposium on VLSI (GLSVLSI)*, Alberta, 2017.
- Michail Noltsis, D. Rodopoulos, N. Zompakis, F. Catthoor, and D. Soudris. 2017. "Runtime Slack Creation for Processor Performance Variability using System

Scenarios", in *ACM Trans. Des. Autom. Electron Syst. (TODAES)*, Article 24, December 2017.
- Michail Noltsis, N. Zambelis, F. Catthoor, and D. Soudris, "A Closed-Loop Controller to Ensure Performance and Temperature Constraints for Dynamic Applications", in *ACM Trans. Embed. Comput. Syst. (TECS)*, Article 40, October 2019.
- J.-Y. Lin, P. Weckx, S. Mishra, A. Spessot, and F. Catthoor, "Proactive run-time mitigation for time-critical applications using dynamic scenario methodology," in *2022 Design, Automation & Test in Europe Conference & Exhibition (DATE)*, pp. 616–621, 2022.
- J.-Y. Lin, P. Weckx, S. Mishra, A. Spessot, and F. Catthoor, "Multitimescale mitigation for performance variability improvement in time-critical systems," *IEEE Transactions on Very Large Scale Integration (VLSI) Systems*, vol. 30, no. 11, pp. 1757–1769, 2022.
- B. Ranjbar, F. Klemme, P. R. Genssler, H. Amrouch, J. Jung, S. Dave, H. So K. Lee, A. Shrivastava, J.-Y. Lin, P. Weckx, S. Mishra, F. Catthoor, D. Biswas, and A. Kumar, "Learning-Oriented Reliability Improvement of Computing Systems From Transistor to Application Level," in *2023 Design, Automation & Test in Europe Conference & Exhibition (DATE)*, April 2023.

9.4 Perspective

In general, our research serves as the preliminary steps beyond the conventional WCET-based DVFS scheduling approach for ensuring timing guarantees. There is plenty of room left for extending our work to an all-around solution. Therefore, we identified the important continuation and extension of this work as follows:

- Chapter 4 deals with PID implementation using the system and adaptive scenarios. In the former implementation, we used the GSM spectrum sensing application and performed an offline analysis that clustered its nearly 800 RTSs into 30 scenarios. An exploration of different clustering profiles (such as the use of 5, 10, or 20 scenarios) and the study of the response of our scheme in these cases is definitely interesting. Such a work could provide significant information regarding the trade-off between the complexity of the scenario analysis and the response of the PID controller.
- The dynamic scenario principles described in Chap. 5 can be potentially instantiated on more complicated architectures, such as multicore processors and GPU, to develop architectural-specific mitigation approaches for performance variability. Since these architectures are inherently equipped with multiple functional units of various types, the area cost and design spaces of the knobs could even be smaller than single-core processors. In reality, there is also a need to ensure timing guarantees for these architectures that run complex workloads, such as car auto-driving systems. Hence, further research is required to develop mitigation approaches for these systems.

- The HDP knob proposed in Chap. 5 is an instantiation of how the inherent performance-energy trade-off in microarchitecture can be exploited to realize a fast-switching system knob. We suppose there is more potential for designing other microarchitectural knobs on this principle in processor cores, memory, and interconnect networks. In addition, supports of EDA tools are also crucial for optimizing the energy efficiency of these knobs. Therefore, we look forward to the extension of the concept of the microarchitectural knobs.
- We have demonstrated the dynamic scenario-based algorithm for timing guarantees of a job sequence in Chap. 5. There is plenty of room for generalizing the algorithms for various purposes. For example, adapting the algorithm to the constraints of maximum deadline misses or maximum latency could make it suitable for time-sensitive and general applications. Also, by integrating the algorithm with real-time scheduling algorithms, a timing-guaranteed solution for multiple tasks in multicore processors can be realized.
- The cycle-accurate simulation flow in Chap. 6 can be extended to simulate the performance variability and mitigation of more sophisticated architectures, such as including multilevel caches, multicore processors, system-on-chip, etc. However, huge computation resource is required for performing this kind of simulation, which is a challenge to overcome.
- In Chap. 7, the proposed cross-layer reliability and variability mitigation framework covers the levels from the circuit to the processor. Beyond our framework, the higher levels (e.g., macroarchitecture, software) are involved with very complex systems with a large amount of virtualization and dynamism. Therefore, they are also relevant to explore in order to reduce the complexity to ensure timing guarantees.
- The cross-layer analysis in Chap. 8 reveals how the requirements for functional correctness and timing guarantees can affect each other. We speculate that similar relations exist among other reliability objectives. Therefore, we recommend both theoretical and empirical studies to understand the interdependencies of all objectives (functional/parametric correctness, timing, thermal, lifetime), which will be a crucial step in the realization of multi-objective reliability solutions.
- The cross-layer reliability and variability mitigation framework we proposed in Chaps. 7 and 8 is suitable for providing guarantees for the constraints we target, namely timing and functional correctness. In a real-world system, there are more QoS constraints that should be met at the same time, such as chip area and thermal constraints. One should be able to extend our framework on the same principles of dynamic scenarios, while an N-dimensional Pareto space of QoS constraints is incorporated. In this way, a reliability solution that ensures guarantees of N QoS constraints can be realized.

In conclusion, while the aforementioned gaps are required to be filled, we look forward to the development of a general performance variability mitigation approach, which can ensure timing guarantees for highly reliable products.

Index

J.-Y. Lin et al., *Workload Dependent Mitigation Approaches for Performance Variability*, https://doi.org/10.1007/978-3-031-85624-2

The manufacturer's authorised representative in the EU is Springer Nature Customer Service Centre GmbH, Europaplatz 3, 69115 Heidelberg, Germany. If you have any concerns regarding our products, please contact ProductSafety@springernature.com

Printed and bound by CPI Group (UK) Ltd, Croydon, CR0 4YY

01/10/2026

02225956-0001